北京大学专业英语丛书

化学专业基础英语 I
（第2版）

INTRODUCTORY CHEMISTRY SPECIALITY ENGLISH (2nd Edition)

魏高原 编
by Gaoyuan Wei

北京大学出版社
PEKING UNIVERSITY PRESS

图书在版编目(CIP)数据

化学专业基础英语(Ⅰ)/魏高原编. —2版. —北京：北京大学出版社，2012.7
(北京大学专业英语丛书)
ISBN 978-7-301-20883-0

Ⅰ.①化…　Ⅱ.①魏…　Ⅲ.①化学－英语－高等学校－教材　Ⅳ.①H31

中国版本图书馆 CIP 数据核字(2012)第 139797 号

书　　名：化学专业基础英语(Ⅰ)(第2版)
著作责任者：魏高原　编
责 任 编 辑：赵学范
标 准 书 号：ISBN 978-7-301-20883-0/O·0874
出 版 发 行：北京大学出版社
地　　址：北京市海淀区成府路205号　100871
网　　址：http://www.pup.cn　电子信箱：zpup@pup.pku.edu.cn
电　　话：邮购部 62752015　发行部 62750672　编辑部 62767347　出版部 62754962
印 刷 者：三河市博文印刷有限公司
经 销 者：新华书店
　　　　　787 毫米×1092 毫米　16 开本　17 印张　500 千字
　　　　　2001 年 4 月第 1 版
　　　　　2012 年 7 月第 2 版　2023 年 7 月第12次印刷(总第 25 次印刷)
印　　数：92501～95500 册
定　　价：42.00 元

未经许可，不得以任何方式复制或抄袭本书之部分或全部内容
版权所有，侵权必究
举报电话：(010)62752024　电子信箱：fd@pup.pku.edu.cn

内 容 简 介

本教材第 2 版根据北京大学化学学院试用多年的讲义及第 1 版教材修订而成，无论从内容取舍还是从教学目的上看都是有开创性的。本书在内容编排上试图训练学生在系统掌握专业英语词汇的基础上，学会用英语进行科学思维。经北大化学院多年教学使用，受到学生普遍欢迎。全书 50 万字，分成基础化学讲座、重要专业术语和化学文献选讲及附录四部分。附录中为读者提供了习题答案和试题、基本化学术语总汇以及一些阅读、会话、写作和翻译用资料及提高听力用的化学录像、光盘及计算机课件目录。第 2 版的内容更加丰富。

本书可与《化学专业基础英语（Ⅱ）》配套使用——在接受本书的系统学习同时（或之后），如配合基础英语（Ⅱ）的学习，读者则不难在领略多彩的化学世界前沿领域的同时，全面提高专业英语水平。

本书可作为大专院校化学及相关专业高年级学生专业英语教材或主要参考书，也可作为理工类研究生和教师以及一般科研人员的实用科技英语参考读物。

序
(Preface)

《化学专业基础英语(Ⅰ)》是根据教育部批准的《大学英语教学大纲》关于专业英语教学的要求和编者本人多年化学专业英语教学实践而编写的一部教材。该教材已在北京大学化学学院试用过七个学期,受到学生普遍欢迎。

该教材的目的是培养学习者在化学专业英语方面的较强的读、写、译的能力,并适当训练听、说的能力,不是单纯培养阅读能力。教材分三个部分:基础化学讲座、重要专业术语、化学文献选讲。附录部分包括总词汇表,习题答案,会话材料,翻译材料,常见化学计量单位、常数等的英文表达方式,常见科技英语语法结构等。

该教材的文章全部选自原文材料,有一定难度。文章内容虽然是有关化学,但不枯燥,文字优雅。教材覆盖化学专业所必需的基础知识及重要词汇和语法现象,并突出化学专业英语文献在文章结构、文字表达方面的特点。这一切都有利于学习者提高英语、迅速掌握化学专业英语。

该教材的练习形式新颖,突出实用,使学习者既能获得必要的化学知识以及解决问题的能力,又能发表个人独立的观点,在读、写、译方面得到训练,并且能通过对文章和议题的确切理解和对观点的独立、精确的表达,培养科学精神。具体说来,编者希望通过使用本教材后,学习者能不借助字典读懂内容不十分专业的科技期刊如 *Nature* 和 *Science* 的文章,能写出一篇科技文章的摘要及小论文或能与同行进行有效的书信往来,能听懂一般化学方面的演讲或讲课,能与外国同行进行化学专业方面的口头交流,并能胜任专业知识方面的中英双向口译和笔译。

该教材设有考试样题。练习题和考试题大部分选自美国著名大学同类专业的教材,对学习者是一种挑战。练习题和考试题都附有答案。

该教材可作为大学化学专业英语教材,也可作为化学类基础课的英文教材。该教材可安排在三年级,用一个学期(18周,周学时2)教完。教学重点应是培养较强的化学专业英语的读、写、译的能力。参照教育部《大学英语教学大纲》关于专业英语教学各项指标要求,在使用该教材的过程中,教师可针对学习者的情况,采用灵活的教学方式,在有条件的院校,可用全英语授课,给学习者全面的化学英语训练。

该教材的后续教材《化学专业基础英语(Ⅱ)》选用了美国普林斯顿大学出版社的原版书《设计分子世界——化学前沿》(作者 Philip Ball)。各章配有内容提要、词汇表、难句解释和翻译,使学习者进一步提高化学专业英语水平,并了解当今化学领域的新成果、新思想。

安美华

北京大学英语系

2001.3.25

编者的话(第 2 版)
(Words from the Editor for the 2nd Edition)

本教材自第 1 版于 2001 年 4 月出版以来,一直得到使用本书的大专院校专业英语教师及其他行业读者的厚爱。他们提出很多非常好的纠错与改进建议,包括河南省许昌学院"化学专业英语课件"课题组全体成员。后者还曾邀请本人去他们学院共商将本书做成多媒体教学课件事宜。当然,对本教材的改进提出最多也是最好的改进建议的当属本人所教的十多届北京大学化学学院本科生。特别是北京大学外国语学院英语系安美华教授给第 1 版所写的前言,对于本书的广泛传播起到了很大的促进作用。本书已经印刷了十多次,不能不促使本人于百忙中抽出时间尽快修订第 1 版。经过本人的修订及本书责任编辑赵学范编审的积极建议和认真审读,终于使本书第 2 版在篇章结构、体例和版式等诸方面均取得了令人满意的效果。

根据本人在北京大学使用本教材的教学经验,本打算在第 2 版里增加专业英语的写作与翻译方面的教学内容。但考虑到这样会显著增加本书的篇幅及售价,最终还是放弃了。好在过去十余年里已经出版了不少这方面的教材,例如田传茂主编的《大学科技英语》(湖北科学技术出版社,武汉,2007),魏汝尧与董益坤主编的《科技英语教程》及《科技英语教程学习指导》(北京大学出版社,北京,2005)及吴炯圻编著的《数学专业英语(第 2 版)》(普通高等教育"十一五"国家级规划教材,高等教育出版社,北京,2009)里就有关于科技英语的理解、翻译和写作的很好的阐述与许多应用实例,以及有关科技(例如数学)论文写作及英语文献查阅的介绍。需要这方面教学内容的教师可以考虑将之与第 2 版《化学专业基础英语(Ⅰ)》一起使用。此外,本人在过去的课堂教学中还广泛使用了许多化学视频来训练学生专业英语听力,特别是美国进口原版化学电影节目《The Super-charged World of Chemistry(超酷化学世界)》(Standard Deviants Video Course Review, Debbie Mintz, et al., Cerebellum Corp., 1996)效果甚佳,很受学生欢迎。作为化学专业英语课外阅读补充读物,还可使用以下两本教材:刘宇红主编的《化学化工专业英语》(中国轻工业出版社,北京,2000)和由英国 Micheael Lewis 编写及荣国斌和铙腊霞注释的牛津专业英语基础丛书《Advanced Chemistry through Diagrams(化学专业英语基础——图示教程)》(上海外语教育出版社,上海,2000)。后者为牛津大学出版的英国 A-Level(相当于大学预科)考试复习用书,以图表形式归纳整理了化学学科的主要知识。

此次再版主要删减了第 1 版"附录 G 常用科技英语词汇(Word Study Material)"中的部分练习题以及"附录 I 基本化学术语总汇(Basic Chemical Terms)"的第一部分,增加了两章有关生物化学和高分子化学的基础讲座(第 8 章和第 9 章),以及"环境化学专业术语"(第 16 章)和"药物化学专业术语"(第 17 章)两章。在已有专业术语章节里,"有机化学专业术语"里增加了两节有关有机化合物基础知识方面的内容,"高分子化学专业术语"里增加了聚合物回收利用方面的内容,而"生物化学专业术语"里则增加了有关 DNA 与 RNA 等生物大分子在自然界起源方面的内容。在"化学文献选讲"部分增加了一篇发表在*Nature*(《自然》)杂志上的介绍神奇拉胀分子网络的分子力学模拟计算的研究论文。最后,重新编写了附录 A 中的表

A.7(Table A.7 List of Audio-Video Material Teaching Basic Chemistry),增加了化学方面的影像资料和多媒体课件等内容,并对第 1 版附录 B～J 重新予以编排,特别是在"基本化学术语总汇"里增加了索引页码。

最后,非常感谢下列化学文献的作者和出版商的支持,使得本教材的再版得以顺利完成。它们是:

(1) *An Introduction to Chemistry* (Mark Bishop, Pearson Edcuation, Inc., San Francisco, CA, USA. 1st ed., 2002);

(2)《环境化学——*Chemistry of the Environment*》(大学环境教育丛书,【美】斯派罗,斯蒂格利亚尼著,影印本,北京:清华大学出版社,第 2 版,2003)及其英文原著 *Chemistry of the Environment* (Thomas G. Spiro and William M. Stigliani, Pearson Education Asia Ltd., 2nd ed., 2003);

(3) 网页 http://en.wikipedia.org/wiki/medicinal chemistry 和 http://www.chem.qmul.ac.uk/iupac/medchem;

(4)《化学专业基础英语(Ⅱ)(设计分子世界:化学前沿)》(北京大学专业英语丛书,【美】Philip Ball 著,魏高原,王剑波,甘良兵 注释,北京大学出版社,2001)及其英文原著 *Designing the Molecular World: Chemistry at the Frontier* (Philip Ball, Princeton University Press, Princeton, New Jersey, USA, 1994)。

魏高原

2011.9.28

于北京大学

编者的话(第 1 版)
(Words from the Editor for the 1st Edition)

现代通信和运输工具正使"地球村"这一设想日益变成现实,而同村的人必须能够进行有效的语言交流。尽管具有悠久历史的象形文字——中国汉字在计算机技术出现之前,一直难以为非华语使用者广泛应用,但相信在不久的将来,必定会有更多的地球村人能够使用这一令中华民族引以为豪的语言文字。不过,由于众所周知的原因,可以预见在未来二三十年内,英语仍将作为国际交往中使用最普遍的一种语言,特别对科学技术领域更是如此。更考虑到落实"科教兴国"战略的需要,并且教育部又于 1999 年 6 月颁布了新的《大学英语教学大纲》,新大纲明确规定:"学生在完成基础阶段的学习任务,达到英语四级或六级后,都必须修读专业英语,以便从学习阶段过渡到应用阶段"。新大纲还将专业英语定为必修课,要求教学学时数不少于 100 学时。此外,大纲对应用提高阶段中在词汇、读、听、说、写、译等方面提出了具体要求。本教材的编写正是在此大背景下应运而生,期望能在起到抛砖引玉作用的同时,缓解目前高校新型化学专业英语教材紧缺这一燃眉之急。

本教材是在 1993 年秋由编者完成的北京大学化学系"化学专业英语"课讲义《ENGLISH FOR CHEMISTRY STUDENTS——LECTURE NOTES》(胶印版)基础上整理、增补而成。新教材保留了原教材的风格,即突出对学生用英语进行科学思考的能力的训练。全书共 16 章,分成以下三大部分内容:基础化学讲座、重要专业术语和化学文献选讲。第一部分(第 1~7 章)内容的安排在国内抑或在全世界尚属首次,这主要是考虑到了编者本人在专业英语学习过程中所积累的一些经验。特别是每章后面所附普通化学习题练习(家庭作业)更系编者本人在国外攻读学位期间应用过的一种行之有效的学习专业英语的方法。同时,所教过的学生对此部分的内容普遍表示欢迎。第二部分(第 8~13 章)主要是为了扩大学生的专业词汇,以及训练学生准确理解专业术语精确定义的能力。第三部分(第 14~16 章)提供了若干有代表性的化学专业文献供选读,主要目的是让学生获得快速理解不同类型专业文献的技巧。最后,特别值得一提的是,本教材还在第四部分附录中为读者提供了较多的参考和补充资料,特别是在"基本化学术语总汇"中列出了本教材中出现的所有基础专业术语,相信会对读者在化学专业术语的掌握方面进行自我测试带来方便。此外,专业英语电化教学方面的参考资料在附录中有所提及,但限于篇幅,未能提供更多资料,希望将来能有这方面的专门教材问世。鉴于本教材属化学专业英语的入门教材,若能与同由北京大学出版社出版的《化学专业基础英语(Ⅱ)——设计分子世界:化学前沿》(已列入《北京大学专业英语丛书》)联用,则效果会更好。

尽管本教材的主要内容已经在北京大学化学与分子工程学院讲授了 7 个学期,但由于属首次尝试,再加上时间仓促,错漏等不完善之处在所难免,敬请读者不吝指正。

本书能以今天的面貌出现,是与众多领导、师生和亲友的支持和帮助分不开的。编者特别感谢原化学系主管教学的副系主任常文保教授在过去几年里从各方面所给予的支持、鼓励和帮助,以及北京大学出版社领导和本书责任编辑赵学范老师在为使本书得以如期出版方面所给予的大力支持和帮助。责编的高度敬业精神和高超编辑水平十分令人感动和钦佩。北京大学英语系安美华教授在使编者学会如何教好专业英语方面给予了很多宝贵的指点。化学学院的同行也给予编者很大帮助,这里要特别提到的有张榕森、甘良兵和王剑波,后者还试用过编者编写的胶印版讲义,并提出过宝贵的改进建议。还有编者所教过的数百名本科生的宝贵批评和鼓励意见,更是编者坚持将此教材完成的强大驱动力。最后,编者感谢夫人在过去几年里所给予的支持和谅解!

在本教材的编写过程中,编者引用了众多科技英语、期刊、教材、专著等英文原版参考文献中的有用部分。这些文献包括:*Chemistry*(2nd ed., John C. Bailar, Jr. et. al., Academic Press, Orlando, Florida, 1984——本教材的主要参考书)、*Fundamentals of Analytical Chemistry*(Douglas. A. Skoog and Donald M. West, Holt, Rinehart and Winston, Inc., 1963)、*Polymer Chemistry:The Basic Concepts*(Paul C. Hiemenz, Marcel Dekker, Inc., New York, 1984)、*Principles of Polymer Chemistry*(Paul J. Flory, Cornell University Press, Ithaca, New York, 1953)、*Biophysical Chemistry Part I:The Conformation of Biological Macromolecules*(Charles R. Cantor and Paul R. Schimmel, W. H. Freeman and Co., San Francisco, 1980)、*Scientifically Speaking:An Introduction to the English of Science and Technology*(B. C. Brookes, Bob Kesten, Viola Huggins, B. B. C. English by Radio and Television & The Chaucer Press, UK, 1971)和 *A Course in Basic Scientific English*(J. R. Ewer and G. Latorre, Longman Group Ltd, London, 1969 & 1976)。在此,对作者和出版商们的支持表示衷心的感谢。

魏高原

1999.6.26

于燕园

目 录
CONTENTS

第一部分（Part Ⅰ） 基础化学讲座（Chemistry Lectures）

第 1 章（Chapter 1） 化学的本质（The Nature of Chemistry） ……………… (3)
    Homework No. 1 ……………………………………………………… (8)

第 2 章（Chapter 2） 作为定量科学和物质科学的化学（Chemistry as a Quantitative Science and a Science of Matter） …………… (9)
    2.1 Introduction ……………………………………………………… (9)
    2.2 Numbers in Physical Quantities ………………………………… (10)
    2.3 Units of Measurement …………………………………………… (11)
    2.4 The Dimensional Method and Problem Solving ……………… (13)
    2.5 Atoms and Elements ……………………………………………… (15)
    2.6 Atomic Structure: Five Classic Experiments …………………… (15)
    2.7 Nuclear Arithmetic ……………………………………………… (16)
    2.8 Kinds of Matter …………………………………………………… (17)
    Homework No. 2 ……………………………………………………… (17)

第 3 章（Chapter 3） 原子、分子和离子（Atoms, Molecules and Ions） ……… (19)
    3.1 Atoms and Ions in Combination ………………………………… (19)
    3.2 Atomic, Molecular and Molar Mass Relationships …………… (22)
    3.3 Composition of a Chemical Compound, Simplest and Empirical Formulas, and Molecular Formulas ………… (23)
    Homework No. 3 ……………………………………………………… (24)

第 4 章（Chapter 4） 气态（The Gaseous State） ………………………………… (26)
    4.1 The Nature of Gases ……………………………………………… (26)
    4.2 Volume, Pressure and Temperature Relationships …………… (26)
    4.3 Mass, Molecular and Molar Relationships …………………… (27)
    4.4 Behavior of Gas Molecules ……………………………………… (29)
    Homework No. 4 ……………………………………………………… (30)

第 5 章（Chapter 5） 化学反应和化学计算法（Chemical Reactions and Stoichiometry） ……………………………………………… (32)
    5.1 Chemical Change: Equations and Types of Reactions ………… (32)
    5.2 Stoichiometry …………………………………………………… (35)
    Homework No. 5 ……………………………………………………… (38)

Ⅰ

第6章(Chapter 6) 热化学(Thermochemistry) (40)
    6.1 Energy (40)
    6.2 Heats of Reaction and Other Enthalpy Changes (41)
    6.3 Measuring Heat (43)
    Homework No. 6 (43)

第7章(Chapter 7) 有机化合物和基团的命名(Nomenclature for Organic Compounds and Groups) (46)
    7.1 Saturated and Unsaturated Hydrocarbons (46)
    7.2 Functional Groups with Covalent Single Bonds (48)
    7.3 Functional Groups with Covalent Double Bonds (49)
    Homework No. 7 (50)

第8章(Chapter 8) 碳水化合物、氨基酸、蛋白质和脂肪(Carbohydrates, Amino Acids, Protein and Fat) (52)
    8.1 Carbohydrates (52)
    8.2 Amino Acids and Protein (56)
    8.3 Fat (61)
    Homework No. 8 (63)

第9章(Chapter 9) 聚合反应：由简单分子到高分子(Polymerization Reaction: from Simple Molecules to Polymers) (67)
    9.1 Step-Growth Polymerization and Condensation Polymers (67)
    9.2 Chain-Growth Polymerization and Addition Polymers (69)
    Homework No. 9 (71)

第二部分(Part II) 重要专业术语(Significant Terms)

第10章(Chapter 10) 无机化学术语(Inorganic Chemical Terms) (75)
第11章(Chapter 11) 有机化学术语(Organic Chemical Terms) (86)
    11.1 Hydrocarbons (86)
    11.2 Hydrocarbons and Energy (86)
    11.3 Some Introductory Concepts (87)
    11.4 Organic Compounds and Their Formulas (87)
    11.5 Organic Chemistry as a Basis for Studying Biochemisty and Synthetic Polymers (89)

第12章(Chapter 12) 物理化学术语(Physical Chemical Terms) (90)
第13章(Chapter 13) 分析化学术语(Analytical Chemical Terms) (97)
第14章(Chapter 14) 高分子化学术语(Polymer Chemical Terms) (106)
    14.1 Polymers (106)
    14.2 Natural Polymers (106)
    14.3 Polymer Nomenclature (107)
    14.4 Historical Introduction (109)

  14.5 Recycling Synthetic Polymers ………………………………… (113)

第15章(Chapter 15) 生物化学术语(Biochemical Terms) ……………… (116)
  15.1 Biochemistry ……………………………………………………… (116)
  15.2 Levels of Structures in Biological Macromolecules ………… (118)
  15.3 DNA and Proteins: Which Came First in the Life's
     Chemical Evolution? …………………………………………… (122)
  15.4 The RNA World ………………………………………………… (124)

第16章(Chapter 16) 环境化学术语(Environmental Chemical Terms) …… (126)
  16.1 Prologue on Energy and Sustainability ………………………… (126)
  16.2 Natural Energy Flows …………………………………………… (127)
  16.3 Crisis of Atmospheric Chemistry ……………………………… (129)
  16.4 The Chemical Control of Climate: the Global Warmers …… (130)
  16.5 Green Chemistry and Development of Green
     Chemicals and Catalysts ……………………………………… (131)

第17章(Chapter 17) 药物化学术语(Medicinal Chemical Terms) ……… (134)
  17.1 Scope of Medicinal Chemistry ………………………………… (134)
  17.2 Basic Medicinal Chemical Terms from
     IUPAC Recommendations(1998) ……………………………… (135)
  17.3 Rehabilitation of Old Drugs and Development of New Ones …… (146)

第三部分(Part Ⅲ) 化学文献选讲(Chemical Literature)

第18章(Chapter 18) 说明性短文(Descriptive Short Articles) ………… (151)
  18.1 Laboratory Safety Rules ………………………………………… (151)
  18.2 Keeping Records ………………………………………………… (152)
  18.3 Use of a Balance ………………………………………………… (153)
  18.4 An Experiment …………………………………………………… (155)
  18.5 Molecular Weight of a Substance ……………………………… (156)
  18.6 Tools of Chemistry: Nuclear Magnetic Resonance ………… (159)
  18.7 Tools of Chemistry: the Computer …………………………… (161)

第19章(Chapter 19) 期刊论文(Periodical Papers) ……………………… (164)
  19.1 Progesterone from 3-Acetoxybisnor-5-cholenaldehyde and
     3-Ketobisnor-4-cholenaldehyde ………………………………… (164)
  19.2 Exact Shapes of Random Walks in Two Dimensions ……… (166)
  19.3 Modelling of Molecular Networks with
     Negative Poisson's Ratios ……………………………………… (169)

第20章(Chapter 20) 获奖演说(Award-Receiving Speeches) …………… (172)
  20.1 The Discovery of Crown Ethers
     by Charles J. Pedersen ………………………………………… (172)

20.2 Concept and Innovation in Polymer Science
by Paul J. Flory ·· (176)
20.3 Scientific Research Moves towards the 21st Century
by Sam Edwards ·· (183)

第四部分(Part Ⅳ) 附录(Appendices)

附录 A 单位、常数等实用资料(Tables of Units, Constants and Other Useful Material) ·· (191)
    Table A.1 Fundamental Constants ···································· (191)
    Table A.2 Commonly Used Prefixes ································ (191)
    Table A.3 Base Units of the International System of Units ········ (192)
    Table A.4 Derived Units of the International System of Units ···· (192)
    Table A.5 Atomic Masses Listed Alphabetically ··················· (194)
    Table A.6 Greek Alphabet ·· (197)
    Table A.7 List of Audio-Video Material Teaching Basic Chemistry ··· (197)

附录 B 补充习题与已有习题答案(Additional Homework and Answers to Existing Homework) ··· (198)
    B.1 Additional Homework ·· (198)
    B.2 Answers to Existing Homework ································ (199)

附录 C 试题举例(Sample Mid-Term and Final Exams) ··················· (206)
    C.1 Sample Mid-Term Exam ·· (206)
    C.2 Sample Final Exam ·· (208)
    C.3 A Complete Sample Exam with Answers ····················· (209)

附录 D 科技阅读和翻译课文(Useful Texts for Scientific Reading and Translations) ··· (214)
    D.1 Selected Speed Reading Texts ································ (214)
    D.2 Texts for English-to-Chinese Translation ··················· (222)
    D.3 Texts for Chinese-to-English Translation ··················· (224)

附录 E 科技会话常用课文与词汇(Useful Texts and Vocabulary for Scientific Conversations) ··· (228)
    E.1 Selected Speaking Texts ·· (228)
    E.2 Useful Words for Scientifically Speaking ··················· (236)

附录 F 基本化学术语总汇及索引(Baisc Chemical Terms and Index) ·········· (249)

第一部分 (Part I)

基础化学讲座
(Chemistry Lectures)

第1章 化学的本质
Chapter 1　The Nature of Chemistry

The following is a letter to a friend from John C. Bailar, Jr., who has been a member of the chemistry department faculty at the University of Illinois for 56 years.

Dear Chris:

This letter is an answer to your questions about just what chemistry is and what chemists do. I'm glad that you asked, for many people have a distorted, or at least superficial, view of what the subject is all about. Whether I can give you a clear picture of it in a letter like this, I am not sure, but I shall try.

You know, of course, that chemistry is one of the physical sciences, along with physics, geology, and astronomy. Closely related, but in a somewhat different category, are the biological sciences, such as botany, physiology, ecology, and genetics. There is no sharp distinction between the two groups of sciences, or between those in either group, for they overlap each other. Often it is difficult to decide whether a specific topic belongs in one area or another. Many important subjects fall within the boundaries of several different disciplines. (Definitions of terms given in boldface type are listed at the end of this letter.)

All of the sciences overlap extensively with chemistry: they depend upon it and, in large measure, are based upon it. By that I mean that chemistry is really a part of all of the natural sciences, and a person cannot go very far in any science without some knowledge of chemistry. It would be possible to be a chemist without much knowledge of astronomy or physiology, but certainly, one could not make great progress in astronomy or physiology without some understanding of chemistry. A knowledge of chemistry is essential in other scientific fields as well. Agriculturists, engineers and medical doctors use chemical concepts constantly.

Chemistry is concerned with the composition of matter and the changes in composition which matter undergoes—in brief, chemistry is the science of matter. Physics is concerned chiefly with energy and with the interactions of matter and energy, including energy in such forms as heat, light, sound, electricity, mechanical energy, and nuclear energy. All changes in the composition of matter either release or absorb energy and for this reason the relationship between chemistry and physics is a most intimate one.

We think of any change in which the composition of matter changes as a chemical change. For example, if you pour vinegar on baking soda in a glass vessel, you will see bubbles of gas escaping and the liquid will become warm as energy is released. When the bubbling stops, you can evaporate the liquid by boiling it, until finally only a white powder remains. But this white powder is not the original baking soda. It is a new substance with new characteristics. For example, it won't give off bubbles if you pour vinegar on it. This new material is different in composition from either of the materials which you originally mixed together. A chemical change has taken place.

By contrast, a physical change does not involve a change in the composition of matter. The melting of ice or the stretching of a rubber band are physical changes. It is often impossible to say whether a particular change is chemical or physical. Happily, it is not usually necessary to make a clear distinction between the two.

You must not assume that in your first course in chemistry you'll learn about the chemistry of the digestion of food or how a mixture of cement and water sets and hardens. These are complex processes, and before one can understand them one must first learn the chemistry of simpler substances. In learning to play the piano, a student does not start with Rachmaninoff's *Prelude in C♯ Minor*. A music student must first learn to play scales, and then simple pieces. It is only after months or years of practice that an individual can play the music of the masters. So it is with chemistry. You must first learn the fundamental principles and something about simple substances such as water and oxygen. A good understanding of the behavior of such substances will then allow you to understand the chemical behavior of more complex materials.

The science of chemistry is so broad that no one can be expert in all of its aspects. It is necessary to study the different branches of chemistry separately, and, if you become a chemist, to specialize in one or two branches of the subject.

Until about 150 years ago, it was believed that inanimate matter and living matter were of entirely different natures and had different origins. The inanimate matter was referred to as "inorganic" (meaning "without life") and the living matter and material derived from living matter were called "organic." However, in 1828, a German chemist named Friedrich Wöhler heated a material which was known to be inorganic and obtained a substance which all chemists recognized to be a product formed in life processes. So the distinction between "inorganic" and "organic" broke down. We still use these terms, but they now have different meanings from those they had in the early days. All living matter contains carbon chemically combined with hydrogen, so the chemistry of chemical compounds of carbon and hydrogen, whatever their origin, is called organic chemistry. Substances that do not contain carbon combined with hydrogen are "inorganic," and their chemistry is called inorganic chemistry. Carbon is very versatile in its behavior and is a key substance in a great many compounds, including most of the compounds essential to life.

There are other branches of chemistry, too. Analytical chemistry is concerned with the detection or identification of what substances are present in a material (qualitative analysis) and how much of each is present (quantitative analysis). Physical chemistry is the application of the methods and theories of physics to the study of chemical changes and the properties of matter. Physical chemistry really forms the foundation for all of the other branches of the subject. Biochemistry, as the name implies, is concerned with the chemistry of the processes that take place in living things.

Inorganic, organic, analytical, physical chemistry, and biochemistry are the main branches of chemistry, but it is possible to combine portions of them, or to elaborate on them in many ways. For example, Bioinorganic chemistry deals with the function of the metals that are present in living matter and that are essential to life. Pharmaceutical chemistry is concerned with drugs: their manufacture, their composition, and their effects upon the body. Clinical chemistry is concerned chiefly with the analysis of blood, urine, and other biological materials. Polymer chemistry deals with the formation and behavior of such substances as rayon, nylon, and rubber. (Some people would include inorganic polymers such as glass and quartz.) Environmental chemistry, of course, deals with the composition of the atmosphere and the purity of water supplies—essentially, with the chemistry of our surroundings. Agricultural chemistry is concerned with fertilizers, pesticides, plant growth, the nutrition of farm animals, and every other chemical topic that is involved in farming.

One more topic should be mentioned. This is chemical engineering, which is concerned with the applications of chemistry on a large scale. Chemical engineers design and operate chemical factories; they deal with the economics of making chemicals on a commercial scale. They are also concerned with such processes as distilling, grinding, and drying materials in large amounts—even the study of the friction of liquids and gases flowing through pipes.

Before you can undertake the study of any of these broad fields of chemistry, you will need to take a course, usually called "General Chemistry," which is the basis for more specialized study. You will quickly learn that general chemistry consists of two interrelated parts: descriptive chemistry and principles of chemistry.

Descriptive chemistry generally deals with the "What...?" questions: What does that substance look like? What happens when it is heated? What happens when an electric current flows through it? What occurs when it is mixed with another specific substance? Chemistry is an experimental science and chemists work with a great many substances. It is important that they know the nature of these substances: their solubility in water or other liquids, their flammability, their toxicity, whether they undergo chemical changes in damp air, and many other characteristics. Sometimes the availability and cost of a substance are also important. The descriptive part of the general chemistry course is concerned chiefly with the behavior of some of the simpler inorganic substances, but often includes brief discussions of organic and biochemical materials as well.

The principles part of the course is concerned with theories of chemical behavior. That is, it attempts to answer the "Why...?" questions: Why won't a substance dissolve in water? Why did an explosion take place when a mixture was heated? Why was a particular substance and not a different one formed in a chemical change? Why does a chemical change speed up dramatically if a tiny amount of something else is added?

The study of chemical principles is of great practical as well as intellectual interest. We can, for example, calculate how much heat is given off when a particular fuel burns, and determine how to speed up or slow down its combustion. When we know why certain substances behave as they do we can often modify their behavior to achieve desirable or useful results.

Chemistry is an experimental science. By this statement, I do not mean that chemists do not have theories about changes in chemical composition—under what conditions they will or will not take place, how they take place, and what the products will be. There are always theories. But theory must always be subject to experiment. If one's theory is not in accordance with carefully executed experiments, then the theory, not the experiment, must be wrong. The theory must then be abandoned or modified. In this regard, chemistry is quite different from the social sciences, such as sociology and economics. People who work in those fields may have theories about the causes of inflation or unemployment or marital unhappiness, and they may carry out experiments to test their theories. But these experiments can never be repeated and checked under the same conditions, for in the act of doing the experiment the conditions have been irretrievably changed. This is true to some extent also in the biological sciences. A pharmacologist may test the effect of a given drug on a mouse and draw some conclusions from what happens to the mouse. But he cannot repeat the experiment with that same mouse, for he cannot be sure that the health of the mouse has not been changed by the first administration of the drug. He can do the experiment with another mouse, but he cannot be sure that the second mouse will respond exactly as the first one did. Chemists are more fortunate; under the same conditions, pure chemicals will always react with each other in exactly the same way. The trick is to be sure that the chemicals are pure and that the conditions of the experiment are exactly the same.

But, you will ask, "Just what do chemists do?" That is a difficult question to answer, for chemists do many different things. About half of the chemists in the United States work in laboratories. Some of them are "quality control" chemists. By a variety of laboratory techniques (some simple and some complex), they analyze or otherwise test materials which are to be used or the products of a chemical factory (be it a drug factory, a food factory, or a steel mill) to ensure that these products are uniform and pure. Some chemists do laboratory research, hoping to discover new chemicals or new uses for known chemicals, or to improve methods of making useful chemicals. Some seek to unearth new principles of chemical behavior, and their activities may range from laboratory work to using only pure mathematics. None of this, of course, is hit-and-miss experimentation. A chemist is always guided by a background of both chemical theory and practical experience, and the broader these are the more successful the chemist will be.

But what about those who do not work in laboratories? Are they still chemists? Indeed they are, though they may combine their chemical activity with some other professional work. Some spend their time looking for new uses and markets for substances that the research chemists have discovered; some are teachers (or divide their time between teaching and research); some become writers of scientific articles for newspapers and magazines.

You may be wondering whether you should study chemistry at all. I hope that you will do so, for as I indicated earlier, a knowledge of chemistry is useful, no matter what profession you follow. If you decide to become a mechanical engineer, you'll need to know something about fuels and alloys and corrosion; if a civil engineer, you must have a knowledge of cement, plaster, steel, and other building materials; if an electrical engineer, you'll need a knowledge of how a battery produces electrical energy, and the changes that take place in it when it is recharged, as well as a knowledge of transistors and lasers. Should you become a medical doctor, you'll be dealing with the most complex chemical plant of all—the human body—and the multitude of chemicals in it. My own son, John, studied chemistry for three years as an undergraduate, but after one year in medical school, he returned to his undergraduate college to take a summer course in physical chemistry, for he had discovered that he needed that extra chemical knowledge in his medical studies.

If you decide to go into agriculture, you'll need to know about fertilizers and pesticides, as well as animal nutrition. Even if you enter some profession that seems to have no connection with chemistry, such as law, you'll find a knowledge of chemistry very useful. Lawyers frequently have to deal with patents that concern chemical inventions. Some members of the U. S. Congress have had extensive chemical training, which gives them a great advantage in discussions of environmental pollution, nuclear energy, the regulations of the Food and Drug Administration, and in other legislation that concerns scientific matters.

The chemical profession is so broad that persons of many different interests and temperaments find satisfaction in it. A person who studies chemistry for very long develops habits of thinking logically and clearly. Once he has accomplished that, he can do almost any sort of work.

I hope that you will enjoy your study of chemistry. I have found it to be a fascinating subject, because of its history, the beauty of its logic, and its multitude of applications.

Sincerely,

The signature of
John C. Bailar, Jr.

Homework No. 1

01 Match each of the significant terms in the left-hand column in the table with the most appropriate definition in the right-hand column.

Significant Terms	Explanation
biological sciences	study of living matter
chemistry	the branch of science that deals with matter, with the changes that matter can undergo, and with the laws that describe these changes
chemical change	a change in the composition of matter
descriptive chemistry	description of the elements and their compounds, their physical states, and how they behave
inorganic chemistry	the chemistry of all of the elements and their compounds, with the exception of compounds of carbon with hydrogen and of their derivatives
matter	everything that has mass and occupies space
organic chemistry	the chemistry of compounds of carbon with hydrogen and of their derivatives
physical sciences	study of natural laws and processes other than those peculiar to living matter
physical change	a change in which the composition of the matter involved is unaltered
principles of chemistry	explanations of chemical facts, for example, by theories and mathematics
qualitative analysis	the identification of substances, often in a mixture
quantitative analysis	the determination of the amounts of substances in a mixture
theory	unifying principle or group of principles that explains a body of facts or phenomena

02 Write a letter to a friend or your parents, telling your chemistry-studying life.

第 2 章 作为定量科学和物质科学的化学
Chapter 2 Chemistry as a Quantitative Science and a Science of Matter

2.1 Introduction

Every object in the world around you can be described in terms of chemistry. Many events that you can see occurring in nature involve chemical changes: the changing color of the leaves in the fall, the transformation of a pond into a swamp, the rusting of iron. Curiosity about what can be observed in the world has led to the study of chemistry.

Let's describe what is seen in one specific chemical change. Two substances are involved. One is a black powdery solid. The other is a colorless liquid that causes irritation if spilled on the skin. If some of the black solid is placed in a container and the liquid slowly added, things happen. The black solid begins to dissolve. The solution that is formed is not black, but very pale green. At the same time, a gas begins to bubble out of the solution. And the air is filled with a terrible smell, like that of rotten eggs.

What a multitude of questions can be asked here. What are these substances? Why did the black solid dissolve? What was formed in its place? How much of the liquid does it take to dissolve all of the black solid? How much of the gas can be produced? How long did the change take? Will events speed up if we heat the mixture? If so, by how much per degree of temperature?

Notice how many of the questions are quantitative ones. Observation and measurement both play vital roles in answering the questions of chemistry. A chemical change is not completely understood until it is understood quantitatively—in terms of measurements and numbers. Our understanding of chemistry is tested by making measurements. If a prediction is made based on what we think we understand, and if the prediction is shown to be correct by obtaining the predicted numbers in a quantitative test, we have greater confidence in our understanding.

In studying chemistry you will be presented with facts accumulated during hundreds of years of observation and measurement. You will also learn how the principles of chemistry are used to explain what has been observed. To test your understanding of chemical principles, you will solve problems, frequently problems that utilize the results of measurements of physical properties.

2.2 Numbers in Physical Quantities

1. Measurement and Significant Figures

(1) The result of measuring a physical property is expressed by a numerical value together with a unit of measurement, for example,

    180 pounds   91 kilograms

(2) Exact numbers are numbers with no uncertainty; they arise by directly counting whole items or by definition. Numbers that result from measurements are never exact. There is always some degree of uncertainty due to experimental errors: limitations of the measuring instrument, variations in how each individual makes measurements, or other conditions of the experiment.

(3) Significant figures in a number include all of the digits that are known with certainty, plus the first digit to the right that has an uncertain value. For example, the uncertainty in the mass of a powder sample, i. e., 3.1267g as read from an "analytical balance" is ±0001g.

(4) Errors in measurement:

(i) Random errors which result from uncontrolled variables in an experiment and affect precision—the reproducibility of the results of a measurement;

(ii) Systematic errors which can be assigned to definite causes and affect accuracy—the closeness to the true result of a measurement or an experiment.

2. Finding the Number of Significant Figures

(1) The number of significant figures is found by counting from left to right, beginning with the first nonzero digit and ending with the digit that has the uncertain value, e. g.,

    454(3)   0.296(3)   7.31(3)   0.00846(3)   10.7(3)
                    1520(3)   1520.(4)

N.B. Zeros at the end of a number given without a decimal point present a problem because they are ambiguous. In general, we recommend that such terminal zeros be assumed to be not significant. The ambiguity is removed if a decimal point is given; then all the zeros preceding the decimal point are significant.

(2) *Ex.* How many significant figures are in the numbers (a) 57, (b) 82.9, (c) 340, (d) 700., (e) 10.000, (f) 0.000002, (g) 0.0402, and (h) 0.04020?

3. Arithmetic Using Significant Figures

(1) Addition and subtraction: Round the answer to the place (before or after the decimal point) with the greatest uncertainty, e. g.,

```
    23.2           53            53
     6.052        600           600.
   139.4          168           168
   ───────       ─────         ─────
   168.652        821           821 (±1)
   168.7(±0.1)                  800(±100)
```

(2) Multiplication and division: Round the answer to the same number of significant figures as in the number with the fewest significant figures, e. g.,

$$23.2 \times 0.1257 = [2.91624] = 2.92$$
$$6 \times 6.35 \text{ g} = 38.1 \text{ g}$$

(3) *Ex*. Perform the following calculation and express the answer to the proper number of significant figures.

$$x = \frac{95.316}{2.303 \times 1.987} \times \left(\frac{1}{298} - \frac{1}{308}\right)$$
$$= [2.269378954]$$

$1/298 = 0.00336 \qquad 1/308 = 0.00325$

$1/298 - 1/308 = 0.00011$

$x = 2.3$

4. Scientific Notation (Exponential Notation)

(1) In standard scientific notation the significant figures of a number are retained in a factor between 1 and 10 and the location of the decimal point is indicated by a power of 10, e. g.,

$$0.0063 = 6.3 \times 10^{-3} \qquad 900\,000\,000 = 9 \times 10^8$$

(2) Arithmetic using scientific notation

```
    99 + 1.23×10³        or        0.099×10³
                                   1.23 ×10³
                                   ─────────
                                   1.329×10³
                                   1.33 ×10³
```

$$\frac{4.3 \times 10^{-2}}{2.01 \times 10^{-1}} = 2.1 \times 10^{-1} = 0.21$$

$$(6.022 \times 10^{23}) \times (4.2 \times 10^3) = 25 \times 10^{26} = 2.5 \times 10^{27}$$

2.3 Units of Measurement

1. Systems of Measurement

(1) The weight of an Englishman = 14 stone (89 kilograms)

      an American = 180 pounds (82 kilograms)

      a Canadian = 91 kilograms

(2) Metric system: devised by the French National Academy of Sciences in 1793.

(3) SI system (for Système International): adopted by the International Bureau of Weights and Measures in 1960, it is a revision and extension of the metric system. Scientists and engineers throughout the world in all disciplines are now being urged to use only the SI system of units.

Table 2.1 SI Base Physical Quantities and Units

Quantity (symbol)	Name of Unit	Abbreviation
Length (l)	meter (or metre)	m
Mass (m)	kilogram	kg
Time (t)	second	s
Electric current (I)	ampere	A
Temperature (T)	kelvin	K
Luminous intensity (I_v)	candela	cd
Amount of substance (n)	mole	mol

Table 2.2 SI Derived Physical Quantities and Units

Quantity (symbol)	Name of Unit (symbol)	Derived Unit
Area (A)	square meter	m^2
Volume (V)	cubic meter	m^3
Density (ρ)	kilogram per cubic meter	kg/m^3
Velocity (u)	meter per second	m/s
Pressure (p)	pascal (Pa)	$kg/(m \cdot s^2)$
Energy (E)	joule (J)	$(kg \cdot m^2)/s^2$
Frequency (ν)	hertz (Hz)	$1/s$
Quantity of electricity (Q)	coulomb (C)	$A \cdot s$
Electromotive force (E)	volt (V)	$(kg \cdot m^2)/(A \cdot s^3)$
Force (F)	newton (N)	$(kg \cdot m)/s^2$

Table 2.3 Prefixes for Multiples and Fractions of SI Units

Decimal Location	Prefix	Prefix Symbol
10^{12}	tera	T
10^9	giga	G
10^6	mega	M
10^3	kilo	k
10^2	hecto	h
10	deka	da
0.1	deci	d
10^{-2}	centi	c
10^{-3}	milli	m
10^{-6}	micro	μ
10^{-9}	nano	n
10^{-12}	pico	p
10^{-15}	femto	f
10^{-18}	atto	a

(4) Equivalence between units (The equivalences marked by "*" are exact):

Length   1 mile = 1.6093 km   1 ft = 0.3048 m
          1 inch = 2.54 cm*   1 Å = 0.1 nm

Volume   1 L = 1 dm^3* = 10^{-3} m^3*   1 qt = 0.94635 L
          1 gal = 3.785 L

Mass   1 metric ton = 10^3 kg   1 lb = 0.45359 kg
          1 oz = 28.350 g

Energy   1 cal = 4.184 J*   1 eV = 1.6022×10^{-19} J
          1 erg = 10^{-7} J*   1 L·atm = 101.325 J*

Pressure   1 bar = 10^{-5} Pa*   1 psi = 6894.76 Pa
          1 atm = 101325 Pa* = 760 Torr*
          1 Torr = 1 mmHg = 133.32 Pa

2. Units of Measurement in Chemistry

Length: 1 Å = 10^{-10} m = 0.1 nm

Volume: 1 mL = 10^{-3} L = 1 cm^3

Mass and weight: Both the SI and metric systems rely on the gram, and the multiples and fractions of the gram, as the units for mass. Strictly speaking, weight should be expressed in units of force. In practice, however, the distinction between weight and mass is often ignored. Expressions such as "weigh out 30 grams of this material", or "How many grams does that sample weigh?" are often used.

Density: g/cm^3 = g/mL

Temperature: There are three temperature scales: the SI scale, measured in kelvin units; the Celsius or centigrade scale, measured in degrees Celsius (°C); and the Fahrenheit scale, measured in degrees Fahrenheit (°F).

$$°C = [(°F+40)(5/9)] - 40$$
$$°F = [(°C+40)(9/5)] - 40$$
$$K = °C + 273.15$$

Heat and energy: 1 cal (calorie) = 4.184 J

Force and pressure: 1 N = 1 (kg·m)/s^2
$$1\ Pa = 1\ N/m^2 = 1\ kg/(m·s^2)$$

The units for pressure include atmospheres (atm), bars (bar), pounds per square inch (psi), torr (Torr), and millimeters of mercury (mmHg).

2.4 The Dimensional Method and Problem Solving

1. The Dimensional Method

(1) The numerical value of a measurement should always be expressed together with the correct unit. In a problem, units are multiplied, divided, and cancelled exactly as numbers would be. If the problem is correctly set up and worked, it should produce an

answer in the correct units.

(2) e.g., $9.0 \text{ V} + 3.29 \text{ V} = 12.3 \text{ V}$

$635 \text{ nm} - 91 \text{ nm} = 544 \text{ nm}$

$6 \text{ L} \times 0.3 \text{ atm} = 2 \text{ L} \cdot \text{atm}$

$(29.0 \text{ cm})^2 = 841 \text{ cm}^2$

$3.0 \text{ cm}/(2.0 \text{ s}) = 1.5 \text{ cm/s} = 1.5 \text{ cm} \cdot \text{s}^{-1}$

$203 \text{ kcal}/(69 \text{ kcal}) = 2.9$

2. Conversion Factors

(1) The conversion of a physical quantity from one unit to another is done with conversion factors derived from the numerical relationship between the two units. Choosing the correct conversion factor allows the cancellation of the unwanted units. Conversion factors or physical constants should include a sufficient number of significant figures so as not to affect the uncertainty of the answer. (If a conversion factor is an exact number it can be treated as having as many significant figures as needed.)

(2) e.g., $(1.2 \text{ eV}) \times [1.6 \times 10^{-19} \text{ J}/(1 \text{ eV})] = 1.9 \times 10^{-19} \text{ J}$

$(99.94 \text{ eV}) \times [1.602 \times 10^{-19} \text{ J}/(1 \text{ eV})] = 1.601 \times 10^{-17} \text{ J}$

3. A Problem-Solving Method

(1) To solve a problem, first make sure that you understand exactly what is known and what is unknown. Then try to figure out how the knowns and unknowns in the problem are connected. Pay special attention to units and conversions. In setting up the problem and solving it, check to see if the answer emerges in the correct units. Make sure to obey the rules for the correct number of significant figures in the answer. Finally, see if the answer seems reasonable.

(2) e.g., A small airplane traveled 128 km in 48 min. What is the speed of the airplane in kilometers per hour?

Step 1. Study the problem and be sure you understand it

　　The unknown: the speed of the airplane in km/h

　　The known: the distance traveled = 128 km

　　　　　　　the time to travel that distance = 48 min.

Step 2. Decide how to solve the problem

　　The connection: Speed is a ratio, distance / time

　　The connection-making: km/min and the conversion

　　　　　　　factor (60 min)/(1 h)

Step 3. Set up and solve the problem

　　$[128 \text{ km}/(48 \text{ min})][60 \text{ min}/(1 \text{ h})] = 160 \text{ km/h}$

Step 4. Check the result

　　(a) Significant figures: correct

　　(b) Units: correct

(c) Is the answer reasonable? Yes.

2.5 Atoms and Elements

1. Elements and Their Symbols

Substances which cannot be broken down chemically into simpler substances have historically been known as elements. Chemical elements are symbolized by one- or two-letter abbreviations derived from their modern names, or in some cases from their old Latin names.

Symbols for Some of the Elements

Aluminum	Al	Carbon	C	Nitrogen	N	Silicon	Si
Arsenic	As	Chlorine	Cl	Oxygen	O	Sulfur	S
Bromine	Br	Chromium	Cr	Phosphorus	P	Uranium	U
Calcium	Ca	Hydrogen	H	Platinum	Pt	Zinc	Zn

Table 2.4 Elements with Symbols Not Based on Their Modern Names

Modern name	Symbol	Derivation of Symbol
Antimony	Sb	Stibium
Copper	Cu	Cuprum
Gold	Au	Aurum
Iron	Fe	Ferrum
Lead	Pb	Plumbum
Mercury	Hg	Hydrargyrum
Potassium	K	Kalium
Silver	Ag	Argentum
Sodium	Na	Natrium
Tin	Sn	Stannum
Tungsten	W	Wolfram*

* The name of an ore.

2. Atoms

All matter is composed of tiny particles called atoms, which are themselves composed of smaller particles. An atom has a dense central core, or nucleus, containing positively charged protons and uncharged neutrons. Much lighter, negatively charged electrons occupy a relatively large space around the nucleus. (If an atom were expanded to the size of one of our largest football stadiums, the nucleus would be about the size of a marble at the center.)

2.6 Atomic Structure: Five Classic Experiments

1. Cathode Rays: The Electron e^- (Joseph John Thomson, in 1897)

When an electrical potential is applied across the two electrodes of a gas-discharge tube, the gas within the tube begins to glow, and if the pressure is low enough cathode rays flow from the negative to the positive electrode. These are now known to be streams of the fundamental, subatomic particles called electrons. The electron is assigned a relative charge of -1 (equal to -1.60×10^{-19} C, as first determined by Millikan, with the mass of 9.11×10^{-28} g or 5.49×10^{-4} u).

2. Canal Rays: The Proton p^+ (Ernest Rutherford, in ca. 1898)

When electrons flow in a gas-discharge tube they leave behind positively charged ions, which can themselves flow from the positive to the negative electrode in the form of canal rays. The positively charged ions of hydrogen are fundamental, subatomic particles called protons. Protons have the same charge as electrons but with a positive sign (a relative charge of $+1$) and a mass of 1.67×10^{-24} g or 1.01 u.

3. α-Particle Scattering: The Nucleus (E. Rutherford)

α-Particles are the nuclei of helium atoms. Some radioactive elements break down spontaneously and emit α-particles at high speeds. When such particles strike a metal foil, some of them are deflected back toward their source. Rutherford concluded from this that the target atoms have a dense, central nucleus in which most of their mass and their positive charge are concentrated.

4. 20 Years Later: The Neutron (James Chadwick, in 1932)

Atomic nuclei contain protons, but protons do not account for all of their mass. Chadwick discovered the neutron, a subatomic particle found in the nuclei of all atoms except ordinary hydrogen. The mass of the neutron is similar to that of the proton, but the neutron has no electrical charge.

5. X-Ray Spectra: Atomic Number (Henry G. J. Moseley, in 1913)

From studies of the X-ray spectra of different elements, Moseley found that the wavelengths produced by each element could be related to a single number corresponding to the number of units of positive charge in its nucleus. This atomic number, Z, is equal to the number of protons in the nucleus.

2.7 Nuclear Arithmetic

1. Atomic Number, Isotopes, and Mass Numbers

The mass number of an element (A) is equal to the number of neutrons (N) and protons (Z) in the nuclei of its atoms ($A = Z + N$). Some elements exist in different forms called isotopes, the atoms of which contain different numbers of neutrons and thus have different mass numbers.

2. Atomic Mass

The actual mass of an atom is its mass in grams. The atomic mass unit (u), a unit of relative atomic mass, is defined as 1/12 the mass of an atom of $^{12}_{6}C$, or 1.6606×10^{-24} g. The masses of isotopes are usually given in atomic mass units. The atomic mass of an element is the average mass (in u) of the atoms in the naturally occurring mixture of isotopes.

2.8 Kinds of Matter

1. Pure Substances and Mixtures

Physical properties are those that can be measured or observed without changing the identity or composition of a substance. Chemical properties can only be observed in chemical reactions, in which the identity of at least one substance is changed. A pure substance always has the same physical and chemical properties and is either an element or a compound. An element is a substance that contains only atoms of the same atomic number. (An atom can be defined as the smallest particle of an element that can participate in a chemical reaction.) A chemical compound is a substance in which atoms of two or more elements are combined in a definite ratio. A mixture contains two or more substances that retain their identities. Any homogeneous mixture of two or more substances is a solution. (The substances in a homogeneous mixture are thoroughly intermingled, and the composition and appearance of the mixture are uniform throughout, while a heterogeneous mixture is a mixture in which the individual components of the mixture remain physically separate and can be seen as separate components, although in some cases a microscope is needed.) The solute—the component present in the smaller amount—is said to be dissolved in the solvent. In an aqueous solution the solvent is water.

2. States of Matter

There are three common states of matter: gaseous, liquid, and solid. Transitions between these are known as changes of state. Not all substances can exist in all three states.

Homework No. 2

01 Perform the following calculations and express the answers in the proper number of significant figures.

(a) $423.1 + 0.256 + 100$ (b) $52.987 + 9.3545 + 6.12$ (c) $14.3920 - 4.4$

(d) $(5183) \times (2.2)$ (e) $14.000/6.1$ (f) $(6.11) \times (\pi)$

(g) $(14.3) \times (60)$ (h) $1020/1.2$ (i) $(3.2) \times (454)/(8.6214)$

(j) $(4/3) \pi (2.16)^3$ (k) $(6.0 + 9.57 + 0.61) \times (1.113)$

(l) $(2.93) \times (14.7) + (1203) \times (0.0296) + (9.38) \times (5.2)$

02 A group of students reported the following measurements for the diameter of a quarter: 2.50 cm, 2.42 cm, 2.43 cm, 2.40 cm, and 2.41 cm. (a) Calculate the class average for the diameter and (b) express the uncertainty in the measurement.

To check the accuracy of the result, the "% error" was calculated.

$$\% \text{ error} = [(\text{experimental value})/(\text{true value}) - 1] \times 100\%$$

Using the true diameter as 2.44 cm, (c) calculate the % error of the class average. (d) Do you think that there was a systematic error in the data?

03 Perform the following calculations and express the answer in standard scientific notation:
  (a) $(6.057 \times 10^3) + 9.35$ (b) $(2.35 \times 10^{-14}) - (7.1 \times 10^{-15})$ (c) $(4.51 \times 10^{-3})/(8.78 \times 10^4)$
  (d) $(1812) \times (1492)/1979$ (e) $[(7.33 \times 10^{-3}) + (4.29 \times 10^1)]/[(5.88 \times 10^{-3}) + (4.29 \times 10^1)]$
  (f) $(5 \text{ km}) \times (14.6)$

04 At what temperature will a Fahrenheit thermometer give (a) the same reading as a Celsius thermometer, (b) a reading that is twice that on the Celsius thermometer, and (c) a reading that is numerically the same but opposite in sign from the Celsius scale?

05 A molecule of palmitic acid has a volume of 110 Å^3. When a drop of the acid is placed on water, the molecules spread out on the surface of the water producing a layer that is one molecule thick. The height of the molecule is 4.6 Å in this layer. Calculate the cross sectional area of the molecule. What area in m^2 will 6.022×10^{23} molecules occupy? The area of 1 Å^2 is equivalent to 10^{-20} m^2.

06 A very important constant that we encounter in this book is known as the ideal gas constant. It is numerically equal to 8.314 J/(K·mol). Express the value of this constant in (a) erg/(K·mol), (b) cal/(K·mol), (c) L·atm/(K·mol).

07 The mass of an empty container is 66.734 g. The mass of the container filled with water is 91.786 g. (a) Calculate the volume of the container using a density of 1.0000 g/cm^3 for water. A piece of metal was added to the empty container and the combined mass was 87.807 g. (b) Calculate the mass of the metal. The container with the metal was filled with water and the mass of the entire system was 105.408 g. (c) What mass of water was added? (d) What volume of water was added? (e) What is the volume of the metal? (f) Calculate the density of the metal.

08 Complete the following table for the atoms or ions as illustrated for $_{20}^{41}\text{Ca}^{2+}$:

Symbol	Z	N	A	Number of Electrons	Electrical Charge
$_{20}^{41}\text{Ca}^{2+}$	20	21	41	18	+2
$_{78}^{190}\text{Pt}$	?	?	?	?	?
$_{53}^{139}\text{I}^-$	?	?	?	?	?
?	14	15	?	?	0
$_?^?\text{Au}^?$	?	?	188	76	?

09 Choose from the following list the symbols that represent (a) groups of isotopes of the same element, (b) atoms with the same number of neutrons, and (c) atoms with the same mass number (4 different sets): (i) ^{12}N, (ii) ^{13}B, (iii) ^{13}N, (iv) ^{14}C, (v) ^{14}N, (vi) ^{15}N, (vii) ^{16}N, (viii) ^{16}O, (ix) ^{17}N, (x) ^{17}F, and (xi) ^{18}Ne.

10 The average atomic mass of chlorine is 35.453 u. There are only two isotopes in naturally occurring chlorine: ^{35}Cl (34.96885 u) and ^{37}Cl (36.96712 u). Calculate the percent composition of naturally occurring chlorine.

第3章 原子、分子和离子
Chapter 3 Atoms, Molecules and Ions

3.1 Atoms and Ions in Combination

1. Molecular and Ionic Compounds

When two or more atoms combine chemically they form a molecule. (A molecule is the smallest particle of a pure substance that has the composition and properties of that substance and is capable of independent existence.) The naturally occurring forms of some elements are diatomic molecules (molecules consisting of two atoms) or polyatomic molecules (which contain more than two atoms). We refer to the compounds composed of molecules as molecular compounds. When an atom gains one or more electrons it acquires a negative charge and is known as an anion; when an atom loses one or more electrons it acquires a positive charge and is known as a cation. An ionic compound (e.g., NaCl) consists of positive and negative ions (Na^+ and Cl^-) held together by electrical attraction. The chemical formula of an ionic compound gives the ratio of ions, but individual molecules are not ordinarily present.

Table 3.1 Chemical Formulas for Molecules of Elements

Monatomic Molecules		Diatomic Molecules		Polyatomic Molecules	
He	Helium	H_2	Hydrogen	P_4	Phosphorus
Ne	Neon	O_2	Oxygen	As_4	Arsenic
Ar	Argon	N_2	Nitrogen	Sb_4	Antimony
Kr	Krypton	F_2	Fluorine	S_8	Sulfur
Xe	Xenon	Cl_2	Chlorine	Se_8	Selenium
Rn	Radon	I_2	Iodine		

Table 3.2 Some Monatomic Ions

H^+, Li^+, Na^+, K^+, Cu^+, Cu^{2+}, Ag^+, Mg^{2+}, Ca^{2+}, Zn^{2+}, Hg^{2+}, Fe^{2+}, Fe^{3+}, Al^{3+}, Bi^{3+}, Cr^{2+}, Cr^{3+}, Co^{2+}, Co^{3+}, Mn^{2+}, Mn^{3+}, Sn^{2+}, Pb^{2+}, F^-, Cl^-, Br^-, I^-, O^{2-}, S^{2-}, N^{3-}, P^{3-}.

2. Formulas for Chemical Compounds

A chemical formula gives the symbols for the elements in a compound with subscripts indicating the number of atoms of each element present. For a molecular compound, the formula represents the number of atoms in one molecule. For an ionic compound, the formula gives the ratio of ions present in the simplest unit, or one formula unit. A structural formula is essentially a diagram showing how the atoms in a compound or ion are linked to each other by chemical bonds. The formula $Mg^{2+}(NO_3^-)_2$ is read "M-G-N-oh-three-taken-twice."

Chemical Formulas for Some Simple Compounds	
Water	H_2O
Carbon monoxide	CO
Carbon dioxide	CO_2
Sulfur dioxide	SO_2
Silver sulfide	Ag_2S
Potassium chloride	KCl
Ammonia	NH_3
Methane	CH_4

Polyatomic Ions			
NH_4^+	Ammonium ion	NO_2^-	Nitrite ion
CN^-	Cyanide ion	NO_3^-	Nitrate ion
CO_3^{2-}	Carbonate ion	O_2^{2-}	Peroxide ion
ClO_3^-	Chlorate ion	OH^-	Hydroxide ion
ClO_4^-	Perchlorate ion	PO_4^{3-}	Phosphate ion
CrO_4^{2-}	Chromate ion	SO_3^{2-}	Sulfite ion
$Cr_2O_7^{2-}$	Dichromate ion	SO_4^{2-}	Sulfate ion
MnO_4^-	Permanganate ion	CH_3COO^-	Acetate ion

Structural formulas:

$$\left[\begin{array}{c} O \\ | \\ O-S-O \\ | \\ O \end{array} \right]^{2-} \quad \text{and} \quad H-\underset{\underset{H}{|}}{\overset{\overset{H}{|}}{C}}-\underset{\underset{H}{|}}{\overset{\overset{H}{|}}{C}}-O-H \quad \text{and} \quad \pentagon$$

3. Naming Chemical Compounds

The rules that govern the naming of chemical compounds are known collectively as chemical nomenclature. In the Stock system, the name of a cation consists of the name of the element, the charge on the ion as a Roman numeral in parentheses, and the word "ion". The name of a monatomic anion (e.g., Cl^-) consists of the name of the element with the ending "ide", followed by the word "ion". A binary compound is one containing atoms or ions of only two elements. Salts are ionic compounds formed between cations and the anions of acids. For binary molecular compounds, prefixes are used to indicate the number of atoms of each element present.

第 3 章 原子、分子和离子(Atoms, Molecules and Ions)

	Older System	Stock System
Mn^{2+}	Manganous ion	Manganese (II) ion
Mn^{3+}	Manganic ion	Manganese (III) ion
Cu^+	Cuprous ion	Copper (I) ion
Cu^{2+}	Cupric ion	Copper (II) ion
Cl^-		Chloride ion
O^{2-}		Oxide ion
N^{3-}		Nitride ion
N_3^-		Azide ion
O_2^{2-}		Peroxide ion
$CuCl$	Cuprous chloride	Copper (I) chloride
$CuCl_2$	Cupric chloride	Copper (II) chloride
Na_3P		Sodium phosphide
$Al_2(SO_4)_3$		Aluminum sulfate

Acid		Anion	
Hydrochloric acid	$HCl(aq)$	Chloride ion	Cl^-
Carbonic acid	$H_2CO_3(aq)$	Carbonate ion	CO_3^{2-}
		Hydrogen carbonate ion	HCO_3^-
Nitric acid	HNO_3	Nitrate ion	NO_3^-
Nitrous acid	$HNO_2(aq)$	Nitrite ion	NO_2^-
Perchloric acid	$HClO_4$	Perchlorate ion	ClO_4^-
Phosphoric acid	H_3PO_4	Phosphate ion	PO_4^{3-}
		Hydrogen phosphate ion	HPO_4^{2-}
		Dihydrogen phosphate ion	$H_2PO_4^-$
Phosphorous acid	H_3PO_3	Hydrogen phosphite ion	HPO_3^{2-}
Sulfuric acid	H_2SO_4	Sulfate ion	SO_4^{2-}
		Hydrogen sulfate ion	HSO_4^-
Sulfurous acid	$H_2SO_3(aq)$	Sulfite ion	SO_3^{2-}
		Hydrogen sulfite ion	HSO_3^-

No. Indicated	1	2	3	4	5	6	7	8	9	10
Prefix	mono	di	tri	tetra	penta	hexa	hepta	octa	nona	deca

N_2O	dinitrogen monoxide	ICl	iodine monochloride
N_2O_5	dinitrogen pentoxide	SO_3	sulfur trioxide

4. Chemical Equations

The substances that undergo changes in a chemical reaction are called the reactants, and the new substances formed are the products. The chemical change that takes place is represented with symbols and formulas in a chemical equation. All chemical equations must be balanced—the correct coefficients must be used for each species so that all the atoms of each element in the reactants can be accounted for in the products. Information about the states of reactants and products may be provided by symbols after the formulas: (g) for gas, (l) for liquid, (s) for solid, and (aq) for substances in aqueous solution. The transformation of a neutral ionic compound into positive and negative ions, usually by dissolution in water, is called dissociation. The formation of ions from a molecular compound is known as ionization. For example,

(1) $$P_4 + 6\ Cl_2 \longrightarrow 4\ PCl_3$$

(read "One P_4 molecule plus six Cl_2 molecules yields four molecules of PCl_3.")

(2) $$N_2(g) + 3\ H_2(g) \xrightarrow{400\ °C,\ 250\ atm,\ FeO} 2\ NH_3(g)$$

(read "Gaseous nitrogen reacts with gaseous hydrogen at 400°C and 250 atm pressure in the presence of FeO as a catalyst to produce gaseous ammonia.")

(3) $$NaCl(s) \xrightarrow{H_2O} Na^+(aq) + Cl^-(aq)$$

("solid sodium chloride; sodium ion in aqueous solution; chloride ion in aqueous solution")

(4) $$HCl(g) \xrightarrow{H_2O} H^+(aq) + Cl^-(aq)$$

("hydrogen chloride; hydrochloric acid")

3.2 Atomic, Molecular and Molar Mass Relationships

1. Molecular Mass

The molecular mass of a chemical compound is the sum of the atomic masses, in atomic mass units, of all the atoms in the formula of the compound. For example,

	number of atoms		atomic mass (u/atom)		mass (u)
N	2	×	14.01	=	28.02
O	5	×	16.00	=	80.00
		molecular mass of N_2O_5	=	108.02 u

2. Avogadro's Number, the Mole, and Molar Mass

Avogadro's number is the number of atoms in exactly 12 g of carbon-12; it is equal to 6.022×10^{23}. A mole is a number of anything equal to Avogadro's number. The mole is the unit that provides the connection between masses on the microscopic level (measured in atomic mass units) and masses on the macroscopic level (measured in grams). The molar mass of a substance is the mass in grams of one mole of that substance.

e.g. 1 How many ozone molecules and how many oxygen atoms are present in 48.00 g of ozone, O_3?

$(48.00 \text{ g}) \times (1 \text{ mol} / 48.00 \text{ g}) \times (6.022 \times 10^{23} \text{ molecules} / 1 \text{ mol})$
$= 6.022 \times 10^{23}$ molecules
$(6.022 \times 10^{23} \text{ molecules}) \times (3 \text{ O atoms} / 1 \text{ molecule})$
$= 1.807 \times 10^{24}$ O atoms

3. Molarity (M): Molar Mass in Solutions

The concentration of a substance in solution is a quantitative statement of the amount of solute in a given amount of solvent or solution. Concentrations are often given in moles per liter of solution, or molarity (M)[①].

e.g. 1 An experiment called for the addition of 1.50 mol of NaOH in the form of a dilute solution. The only sodium hydroxide solution that could be found in the laboratory was a 2 L container marked "0.1035 M NaOH". What volume of this solution would be required for the 1.50 mol of NaOH? If the 2 L container was full, would this be enough?

$(1.50 \text{ mol NaOH})(1 \text{ L} / 0.1035 \text{ mol NaOH}) = 14.5$ L
Not enough

3.3 Composition of a Chemical Compound, Simplest and Empirical Formulas, and Molecular Formulas

The percentage (by mass) of each element present in a chemical compound is its percentage composition. The simplest formula of a compound gives the simplest whole-number ratio of the atoms it contains. An experimentally determined simplest formula is called an empirical formula; it can be determined from the percentage composition and the molar masses of the elements present. The molecular formula of a compound represents the actual number of atoms of each element present in a molecule. To find the molecular formula of a compound it is necessary to know both its empirical formula and its molecular or molar mass, which is usually some multiple of the mass calculated from the empirical formula.

e.g. 1 A sample of potassium metal weighing 3.91 g when burned in oxygen formed a compound weighing 7.11 g and containing only potassium and oxygen. What is the percentage composition of this compound?

$w(K) = (3.91 \text{ g K} / 7.11 \text{ g compound}) \times (100\%) = 55.0\%$
$w(O) = (1 - 0.55) \times (100\%) = 45.0\%$

e.g. 2 The mineral cryolite contains 33% by mass of Na, 13% by mass of Al, and 54% by mass of F. Determine the empirical formula of the compound.

Choose exactly 100 g of cryolite as a basis to solve the problem.

[①] "M" should be written in the SI unit "mol/L".

	Na	Al	F
No. of moles	$\dfrac{33\text{ g}}{22.99\text{ g/mol}} = 1.4$ mol	$\dfrac{13\text{ g}}{26.98\text{ g/mol}} = 0.48$ mol	$\dfrac{54\text{ g}}{19.00\text{ g/mol}} = 2.8$ mol
Mole ratio(n/n_{Al})	$\dfrac{1.4}{0.48} = 2.9$	$\dfrac{0.48}{0.48} = 1$	$\dfrac{2.8}{0.48} = 5.8$
Relative no. of atoms	3	1	6

e. g. 3 The empirical formula for a substance was determined to be CH. The approximate molar mass of the substance was experimentally found to be 79 g. What is the molecular formula of this molecular compound? What is the exact molar mass?

$$[79\text{ g}/\text{mol }(CH)_x] / [(12.01+1.01)\text{g}/\text{mol CH}] = 6.1$$

The molecular formula is $(CH)_6 = C_6H_6$ and the exact molar mass is

$$(13.02\text{ g}/\text{mol}) \times (6) = 78.12\text{ g}/\text{mol}$$

Homework No. 3

01 Write the formula for each of the following simple cations and anions, simple binary compounds, salts of polyatomic ions, and more complex binary molecular compounds:

(a) sodium ion, zinc ion, silver ion, mercury (Ⅱ) ion, iron (Ⅲ) ion, lithium ion, bismuth (Ⅲ) ion, iron (Ⅱ) ion, chromium (Ⅲ) ion, potassium ion, phosphide ion, sulfide ion, telluride ion, chloride ion, and iodide ion;

(b) sodium fluoride, zinc oxide, barium peroxide, magnesium bromide, hydrogen iodide, sodium azide, calcium phosphide, iron (Ⅱ) oxide, silver fluoride, copper (Ⅰ) chloride, potassium azide, manganese (Ⅳ) oxide, and iron (Ⅲ) oxide;

(c) potassium sulfite, calcium permanganate, barium phosphate, copper (Ⅰ) sulfate, ammonium acetate, iron (Ⅱ) perchlorate, potassium nitrite, sodium peroxide, ammonium dichromate, sodium carbonate, silver nitrate, uranium (Ⅳ) sulfate, aluminum acetate, and manganese (Ⅱ) phosphate;

(d) diboron trioxide, silicon dioxide, phosphorus trichloride, sulfur tetrachloride, bromine trifluoride, iodine monobromide, dinitrogen pentasulfide, phosphorus triiodide, silicon monosulfide, and tetrasulfur dinitride.

02 Write a chemical equation representing each of the following reactions: (a) solid aluminum sulfide reacts with liquid water to give solid aluminum hydroxide and gaseous hydrogen sulfide and (b) gaseous ozone reacts with gaseous nitrogen monoxide to produce gaseous nitrogen dioxide and gaseous oxygen. Balance each equation (if you can). For each of the following chemical equations, write a word sentence that describes the chemical reaction

(a) $SiI_4(s) + 2H_2O(l) \xrightarrow{H_2O} SiO_2(s) + 4HI(aq)$

(b) $2H_3AsO_3(aq) + 3H_2S(g) \longrightarrow As_2S_3(s) + 6H_2O(l)$

03 Calculate the molar mass for each of the following: (a) H_3PO_4, (b) $(NH_4)_3AsO_4$, (c) $UO_2(SO_4)$, and (d) $HgBr_2$.

04 You are given a sample containing 0.37 mol of a substance.

(a) How many atoms are present if the sample is uranium metal, U?

(b) How many molecules are present if the sample is acetylene, C_2H_2?

(c) How many formula units are present if the sample is silver chloride, AgCl?

05 What mass of $Al_2(SO_4)_3$ will contain a number of Al^{3+} ions equal to Avogadro's number? What mass of $Al_2(SO_4)_3$ will contain a total number of ions equal to Avogadro's number? What mass of $Al_2(SO_4)_3$ will contain a total number of atoms equal to Avogadro's number?

06 A chemist needed 14.6 g of $CuSO_4$ to perform a chemical reaction.

(a) How many moles of $CuSO_4$ is this? The only source of $CuSO_4$ in the laboratory was a bottle containing $CuSO_4 \cdot 5H_2O$.

(b) How many moles of $CuSO_4 \cdot 5H_2O$ will give the desired amount of $CuSO_4$?

(c) What mass of $CuSO_4 \cdot 5H_2O$ contains 14.6 g of $CuSO_4$?

07 How many moles of acid are present in 108 mL of 0.62 mol/L solution? If we add enough water to make 0.300 L of acid solution, how many moles of acid will it now contain? What is the molarity of the final solution?

08 One type of artificial diamond (commonly called YAG, for yttrium aluminum garnet) can be represented by the formula $Y_3Al_5O_{12}$.

(a) Calculate the percentage composition of this compound.

(b) What is the mass of yttrium present in a 2.0 carat YAG? (1 carat = 200 mg).

09 Calculate the atomic mass of a metal that forms an oxide having the empirical formula M_2O_3 and contains 68.4% of the metal by mass. Identify the metal.

10 A 1.000 g sample of an alcohol was burned in oxygen and produced 1.913 g of CO_2 and 1.174 g of H_2O. The alcohol contained only C, H, and O. What is the empirical formula of the alcohol?

Chapter 4 The Gaseous State

4.1 The Nature of Gases

1. General Properties of Gases

Gases fill whatever space is available to them, but are highly compressible. The pressure of a gas and the space that it occupies vary with temperature. Gases have low densities and flow easily. (The term *volatile* is applied to a substance that forms a vapor very readily.)

CO	Carbon monoxide	Odorless, poisonous
CO_2	Carbon dioxide	Odorless, nonpoisonous
NH_3	Ammonia	Pungent odor, poisonous
PH_3	Phosphine	Terrible odor, very poisonous
CH_4	Methane	Odorless, flammable
C_2H_2	Acetylene	Mild odor, flammable
HCl	Hydrogen chloride	Choking odor, harmful and poisonous
SO_2	Sulfur dioxide	Suffocating odor, irritating to eyes, poisonous
NO_2	Nitrogen dioxide	Red-brown, irritating odor, very poisonous
H_2S	Hydrogen sulfide	Rotten egg odor, very poisonous

2. Kinetic-Molecular Theory of Gases

Molecules of gas are very small in comparison to the spaces that separate them. Gas molecules are in constant random motion, and gas pressure is the result of impacts with the walls. The average kinetic energy of gas molecules is directly proportional to the absolute temperature of the gas. At any given temperature, the molecules of every gas have the same average kinetic energy, molecules of lower mass moving with greater speeds. Collisions with each other and with the walls of the container are perfectly elastic—they involve no loss of energy. A gas having all these properties is known as an ideal gas. The behavior of real gases resembles that of an ideal gas under many conditions.

4.2 Volume, Pressure and Temperature Relationships

1. Variables and Proportionality

A variable quantity is one that can change, for example, the pressure, volume, temperature, and amount of a gas. A constant has a fixed value that cannot change. A and B are directly proportional if $A = kB$, and inversely proportional if $A = k(1/B)$, where k is a constant.

2. Volume Versus Pressure: Boyle's Law

According to Boyle's law, at constant temperature the volume of a given mass of gas is inversely proportional to the pressure upon the gas. For a given mass of gas at constant temperature, $p_1V_1 = p_2V_2$, where the subscripts 1 and 2 designate initial and final conditions.

3. Volume Versus Temperature: Charles' Law

According to Charles' law, at constant pressure the volume of a given mass of gas is directly proportional to the absolute temperature. For a given mass of gas at constant pressure, $V_1/T_1 = V_2/T_2$, where T is the absolute temperature, measured in kelvins (K) on the absolute or kelvin temperature scale, which has a zero point of $-273.15°C$. This is absolute zero, the lowest possible temperature, at which no more kinetic energy can be removed from the molecules of a gas and its volume would in theory fall to zero.

4. p, V, and T Changes in a Fixed Mass of Gas and Standard Temperature and Pressure (STP)

Boyle's law and Charles' law can be combined into a single expression: for a fixed mass of gas, $p_1V_1/T_1 = p_2V_2/T_2$. Gas volumes are generally given at standard temperature and pressure, STP: 0°C (273 K) and 760 Torr (1 atm).

e.g., What is the volume at STP of a sample of gas that occupies 16.5 L at 352°C and 0.275 atm?

$p_1 = 0.275$ atm, $V_1 = 16.5$ L, $T_1 = 352°C + 273 = 625$ K

$p_{STP} = 1.000$ atm, $V_{STP} = ?$ $T_{STP} = 0°C + 273 = 273$ K

$V_{STP} = V_1(p_1/p_{STP})(T_{STP}/T_1) = (16.5 \text{ L}) \times (0.275 \text{ atm} / 1.000 \text{ atm}) \times (273 \text{ K} / 625 \text{ K})$
$= 1.98$ L

4.3 Mass, Molecular and Molar Relationships

1. Gay-Lussac's Law of Combining Volumes and Avogadro's Law

Gay-Lussac observed that when gases react or gaseous products are formed, the ratios of the volumes of the gases involved, measured at the same temperature and pressure, are small whole numbers. The explanation for this law was provided by Avogadro, who recognized that equal volumes of gases, measured at the same temperature and pressure, contain equal numbers of molecules.

e.g. 1 The chemical reaction in laboratory alcohol burners is

$$CH_3CH_2OH \text{ (l)} + 3 O_2 \longrightarrow 2 CO_2 \text{ (g)} + 3 H_2O \text{ (g)}$$
$$\text{ethyl alcohol}$$

What volume of CO_2 (at the same pressure and temperature) is produced if 6.0 L of O_2 reacts?

$$(6.0 \text{ L } O_2) \times (2 \text{ volumes } CO_2 / 3 \text{ volumes } O_2) = 4.0 \text{ L } CO_2$$

e.g. 2 A flask contains 0.116 mol of nitrogen at a given temperature and pressure. Under the same temperature and pressure conditions, a 6.25 ft^3 flask contains 1.522 mol of nitrogen. What is the volume of the first flask?

$$V_1 = n_1(V_2/n_2) = (0.116 \text{ mol}) \times (6.25 \text{ ft}^3 / 1.522 \text{ mol}) = 0.476 \text{ ft}^3$$

2. Molar Volume

Standard molar volume is the volume of one mole of a substance at STP. For any ideal gas, it is equal to 22.4 L. The density of an ideal gas at STP can be found from its molar mass, or vice verse, by using molar volume.

e.g. 1 A pure gas containing 92.3% carbon and 7.7% hydrogen has a density of 1.16 g/L at STP. What is the molecular formula of the gas?

(a) Find the simplest formula.

$$(92.3 \text{ g C}) \times (1 \text{ mol C} / 12.0 \text{ g C}) = 7.69 \text{ mol C}$$
$$(7.7 \text{ g H}) \times (1 \text{ mol H} / 1.0 \text{ g H}) = 7.7 \text{ mol H}$$

(b) Find the molar mass.

$$(1.16 \text{ g/L}) \times (22.4 \text{ L/mol}) = 26.0 \text{ g/mol}$$

(c) Find the molecular formula.

$$(CH)_2 = C_2H_2 \text{ (acetylene)}$$

3. Ideal Gas Law (Mass and Density)

Boyle's law, Charles' law, and Avogadro's law establish relationships for gases among four variables—the volume, the pressure, the temperature, and the amount of gas, which we express in moles, n. When the proportionalities in these laws are combined to give a single relationship

$$pV = \text{constant} \times T \times n$$

and the constant is represented by the symbol R, we have the ideal gas law

$$pV = nRT$$

where R is called the ideal gas constant, the value of which depends on the units used for p, V, and T [i.e. 0.0821 L·atm/(K·mol) and 8.314 J/(K·mol)]. If m is the mass of gas and M is molar mass, $pV = (m/M)RT$. The density, d, of a gas is given by $d = pM/(RT)$.

e.g. 1 A liquid was known to be either methyl alcohol, CH_3OH, or ethyl alcohol, CH_3CH_2OH. The Dumas method for determining molar masses was used to obtain an approximate molar mass, and this value was used to identify the alcohol. The gaseous alcohol at 98°C and 740 Torr had a mass of 0.276 g in a Dumas bulb of volume equal to 270 mL. Which alcohol was present?

$$p = (740 \text{ Torr}) \times (1 \text{ atm} / 760 \text{ Torr}) = 0.97 \text{ atm}$$
$$T = (98+273) \text{ K} = 371 \text{ K}, \quad m = 0.276 \text{ g}$$
$$V = (270 \text{ mL}) \times (1 \text{ L} / 1000 \text{ mL}) = 0.27 \text{ L}$$

$$M = \frac{mRT}{pV} = \frac{(0.276 \text{ g}) \times [0.0821 \text{ L} \cdot \text{atm}/(\text{K} \cdot \text{mol})] \times (371 \text{ K})}{(0.97 \text{ atm}) \times (0.27 \text{ L})}$$
$$= 32 \text{ g/mol}$$

The methyl alcohol is present.

4. Pressure in Gas Mixtures: Dalton's Law

In a mixture of gases that do not react, the molecules of each gas move about independently and distribute themselves uniformly throughout the available space as if no other gas were present. Each therefore exerts the same pressure as it would if it were present alone. This is known as Dalton's law of partial pressure. The pressure of a single gas in a mixture is called its partial pressure.

e.g. 1 A 6.2 L sample of N_2 at 738 Torr is mixed with a 15.2 L sample of O_2 at 325 Torr. The gaseous mixture is placed in a 12.0 L container. What is the pressure of the system?

For N_2: $\quad p_1 = 738$ Torr $\quad V_1 = 6.2$ L

$\qquad\qquad p_2 = ? \qquad\qquad V_2 = 12.0$ L

$\qquad\qquad p_2 = (p_1)(V_1/V_2) = (738 \text{ Torr}) \times (6.2 \text{ L} / 12.0 \text{ L}) = 380$ Torr

For O_2: $\quad p_1 = 325$ Torr $\quad V_1 = 15.2$ L

$\qquad\qquad p_2 = ? \qquad\qquad V_2 = 12.0$ L

$\qquad\qquad p_2 = (325 \text{ Torr}) \times (15.2 \text{ L} / 12.0 \text{ L}) = 412$ Torr

According to Dalton's law of partial pressure,

$$p_{\text{total}} = p_{O_2} + p_{N_2} = 412 \text{ Torr} + 380 \text{ Torr} = 792 \text{ Torr}$$

e.g. 2 What is the composition (in mole fractions) of the gaseous mixture described in e.g. 1?

$x_{N_2} = p_{N_2} / p_{\text{total}} = 380$ Torr $/ 792$ Torr $= 0.48$

$x_{O_2} = p_{O_2} / p_{\text{total}} = 412$ Torr $/ 792$ Torr $= 0.52 \ (= 1 - x_{N_2})$

e.g. 3 A volume of 45.2 mL of "wet" hydrogen gas was collected by the displacement of water at 759.3 Torr and 23.8°C. The vapor pressure of water at this temperature is 22.1 Torr. What mass of "dry" hydrogen gas was collected?

$p = (759.3 \text{ Torr} - 22.1 \text{ Torr})(1 \text{atm} / 760 \text{ Torr}) = 0.9700$ atm

$V = 45.2$ mL $= 0.0452$ L, $M = 2.016$ g/mol, $R = 0.0821$ L $\cdot$ atm/(K $\cdot$ mol)

$T = 23.8$ K $+ 273.15$ K $= 297.0$ K

$$m(H_2) = \frac{MpV}{RT} = \frac{(2.016 \text{ g/mol}) \times (0.9700 \text{ atm}) \times (0.0452 \text{ L})}{0.0821 \text{ L} \cdot \text{atm} \cdot (\text{K} \cdot \text{mol})^{-1} \times (297.0 \text{ K})} = 0.00360 \text{ g}$$

4.4 Behavior of Gas Molecules

1. Effusion and Diffusion: Graham's Laws

Effusion is the escape of gas molecules, one by one, through a hole of molecular dimension. Diffusion is the mixing of different gases by random molecular motions and collisions. According to Graham's laws of effusion and diffusion, at the same temperature and pressure, the rates of both processes are inversely proportional to the square roots of the densities (and therefore to the square roots of the molecular masses) of the gases.

e. g. 1 Find the molar mass of a gas that takes 33.5 s to effuse from a porous container. An identical number of moles of CO_2 takes 25.0 s.

$$M_2/M_1 = (rate_1/rate_2)^2 = (time_2/time_1)^2$$
$$M_2 = M_{CO_2}(time_2/time_{CO_2})^2 = (44.0 \text{ g/mol})(33.5 \text{ s} / 25.0 \text{ s})^2 = 79.0 \text{ g/mol}$$

2. Deviations from the Gas Laws

Molecules of all real gases occupy a finite volume and interact with each other to some extent. These factors cause some deviations from the ideal gas law. The van der Waals equation, which introduces corrections that take into account the volumes and interactions of gas molecules, describes more accurately the behavior of real gases, i.e., $(p + an^2/V^2)(V - nb) = nRT$, in which a is a constant that takes into account the molecular attractions and b is a constant related to the volume actually occupied by the molecules of the substance. Both a nad b are called van der Waals constants. [N. B. (attractions) $\Rightarrow |V_2 - V_1|\_\{real\} > |V_2 - V_1|\_\{ideal\}$ and (excluded volumes) $\Rightarrow |V_2 - V_1|\_\{real\} < |V_2 - V_1|\_\{ideal\}.$]

Homework No. 4

01 More than 34 years ago it was discovered that the "inert" gases are not really inert after all. In particular, xenon reacts with fluorine under various conditions to form a series of compounds.

Colorless crystals of XeF_4 can be prepared by heating a 1 to 5 mixture by volume of Xe to F_2 in a nickel can at 400°C and 6 atm pressure for a few hours and cooling. The equation for the reaction is

$$Xe(g) + 2F_2(g) \longrightarrow XeF_4(s)$$

After the reaction the nickel container contains gaseous F_2 and a little XeF_4 vapor above the crystals of XeF_4.

According to Gay-Lussac's law of combining volumes, (a) what volume of F_2 would react for every milliliter of Xe that reacts? (b) Which law allows us to deduce that the partial pressure of Xe is 1 atm and the partial pressure of F_2 is 5 atm in the original reaction mixture? Predict what will happen to the gases in the nickel container after the reaction if (c) the pressure is increased at constant temperature (use Boyle's law) and (d) the temperature is increased at constant pressure (use Charle's law).

(e) Using Graham's law, predict the increasing order for the rate of effusion of the gases Xe, F_2, and XeF_4.

(f) Basing your argument on actual volumes of molecules, which of the three gases—Xe, F_2 or XeF_4—would you predict would deviate most from ideal gas behavior at high pressures?

02 Calculate the volume of an ideal gas at dry ice (-78.5°C), liquid N_2 (-195.8°C), and liquid He (-268.9°C) temperatures if it occupies 10.00 L at 25.0°C. Assume constant pressure. Plot your results and extrapolate to zero volume. At what temperature is zero volume reached?

03 What temperature would be necessary to double the volume of an ideal gas initially at STP if the pressure decreased by 25%?

04 One liter of PCl_3 in the gaseous state at 200°C and 1 atm reacts with an equal volume of molecular chlorine gas measured under the same temperature and pressure conditions. The product, also a gas, occupies 1 L when measured under the same conditions. What is the formula of this gas?

05 A barge containing 640 tons of liquid chlorine was involved in an accident on the Ohio River. What volume would this amount of chlorine occupy if it were all converted to a gas at 740 Torr and 15°C? Assume that the chlorine is confined to a width of 0.5 mile and an average depth of 50 ft. How long would this chlorine "bubble" be?

06 Cyanogen is 46.2% carbon and 53.8% nitrogen by mass. At a temperature of 25°C and a pressure of 750 Torr, 1.00 g of cyanogen occupies 0.476 L. Determine the empirical formula and the molecular formula of cyanogen.

07 Values of molar mass calculated using the ideal gas law are good only to the extent that the gas behaves as an ideal gas. However, all real gases approach ideal gas behavior at very low pressures, so a common technique for obtaining very accurate molar masses is to measure the density of a gas at various low pressures, calculate d/p from the data, plot d/p against p, extrapolate the curve to $p=0$ to find the intercept, and calculate the molar mass, M, using $M = $ (intercept) RT where $R = 0.0820568$ L·atm/(K·mol). Find the molar mass for SO_2 from the following data at 0°C:

p in atm	0.1	0.01	0.001	0.0001
(d/p) in g/(L·atm)	2.864974	2.858800	2.858183	2.858121

Assume that the temperature and pressure values are exact.

08 The density of dry air at STP is 1.2929 g/L, and that of molecular nitrogen is 1.25055 g/L.

(a) Find the average molar mass of air from these data. The mole fractions of the major components of air are 0.7808 for N_2, 0.2095 for O_2, 0.0093 for Ar, and 0.0003 for CO_2.

(b) Calculate the average molar mass of air.

(c) Compare your answers from (a) and (b).

(d) Why can't Graham's law be used to find the average molar mass of air?

What is the density of (e) dry air at 745 Torr and 25°C and of (f) wet air at 745 Torr and 25°C if it contains water vapor at a partial pressure of 13 Torr?

(g) Explain why the answer to (f) should be less than the answer to (e).

(h) What is the partial pressure of CO_2 in a room containing dry air at 1.00 atm?

(i) What mass of CO_2 will be present at 25°C in a closet 1.5 m by 3.4 m by 3.0 m if the air pressure is 1.00 atm?

(j) What is the predicted pressure of a molar sample of air at -150°C in a 1.00 L container assuming ideal gas behavior?

(k) Repeat the calculation of (j) assuming air to obey the van der Waals gas law with $a = 1.38$ L²·atm/mol² and $b = 0.037$ L/mol.

(l) What is the percentage difference between your answers to (j) and (k)?

第5章 化学反应和化学计算法
Chapter 5 Chemical Reactions and Stoichiometry

5.1 Chemical Change: Equations and Types of Reactions

1. Conservation of Mass and Energy

Two conservation laws apply to all chemical reactions: Energy can neither be created nor destroyed, and matter can neither be created nor destroyed. Thus the atoms taking part in a chemical reaction may be rearranged, but all the atoms present in the reactants must also be present in the products, and the total mass of the reactants must equal the total mass of the products.

e. g. 1

$$2 BaO_2(s) \xrightarrow{\Delta} 2 BaO(s) + O_2(g)$$

barium peroxide   barium oxide   oxygen

(2 mol)(169.3 g/mol)   (2 mol)(153.3 g/mol)   (1 mol)(32.0 g/mol)

338.6 g   =   306.6 g   +   32.0 g

2. Balancing Chemical Equations

A chemical equation must be balanced. That is, it must be written with the correct coefficients for each species participating so that for each element, the number of atoms in the reactants is the same as the number in the products. (To balance an equation, it is often easiest to begin with atoms that appear in only one formula on each side of the equation. It is also best to begin with the most complicated form.)

e. g. 1

$$Al_2S_3(s) + H_2O(l) \xrightarrow{\text{not balanced}} Al(OH)_3(s) + H_2S(g)$$

aluminum sulfide   water   aluminum hydroxide   hydrogen sulfide

Al: $Al_2S_3 + H_2O \xrightarrow{\text{not balanced}} 2Al(OH)_3 + H_2S$

S: $Al_2S_3 + H_2O \xrightarrow{\text{not balanced}} 2Al(OH)_3 + 3H_2S$

O: $Al_2S_3 + 6H_2O \longrightarrow 2Al(OH)_3 + 3H_2S$

e. g. 2

$$Al_2(SO_4)_3(aq) + Ca(OH)_2(aq) \xrightarrow{\text{not balanced}} CaSO_4(s) + Al(OH)_3(s)$$

Al: $Al_2(SO_4)_3 + Ca(OH)_2 \xrightarrow{\text{not balanced}} CaSO_4 + 2Al(OH)_3$

OH: $Al_2(SO_4)_3 + 3Ca(OH)_2 \xrightarrow{\text{not balanced}} CaSO_4 + 2Al(OH)_3$

第 5 章 化学反应和化学计算法 (Chemical Reactions and Stoichiometry)

Ca： $\qquad$ $Al_2(SO_4)_3 + 3Ca(OH)_2 \longrightarrow 3CaSO_4 + 2Al(OH)_3$

e.g. 3

$$C_4H_{10}(g) + O_2(g) \xrightarrow{\text{not balanced}} CO_2(g) + H_2O(g)$$
$$\text{butane}$$

C_4H_{10} (C & H)： $\qquad C_4H_{10} + O_2 \xrightarrow{\text{not balanced}} 4CO_2 + 5H_2O$

O： $\qquad C_4H_{10} + (13/2)O_2 \longrightarrow 4CO_2 + 5H_2O$

$\qquad 2C_4H_{10} + 13O_2 \longrightarrow 8CO_2 + 10H_2O$

3. Some Types of Chemical Reactions

In a combination reaction, two reactants combine to give a single product, i.e., A+B → AB. In a decomposition reaction, a single compound breaks down to give two or more other substances, i.e., AB → A+B. In a displacement reaction, atoms or ions of one substance replace other atoms or ions in a compound (A+BC → AC+B). Metals can be arranged in an activity series based on their ability to displace hydrogen from water or acids and their ability to displace each other in soluble ionic compounds. Partner-exchange (double decomposition, double displacement, and metathesis) reaction have the general form AC + BD → AD + BC. Often such reactions occur between ionic compounds in solution when one product is an insoluble solid, known as a precipitate.

e.g. 1 $\quad 2Li(s) + Cl_2(g) \longrightarrow 2LiCl(s)$
$$\text{lithium chloride}$$

$\qquad NH_3(g) + HCl(g) \longrightarrow NH_4Cl(s)$
$\quad\text{ammonia}\quad\text{hydrogen chloride}\quad\text{ammonium chloride}$

e.g. 2 $\quad 2HgO(s) \xrightarrow{\triangle} 2Hg(l) + O_2(g)$
$\text{mercury (II) oxide}$

$\qquad 2NaHCO_3(s) \xrightarrow{\triangle} Na_2CO_3(s) + H_2O(g) + CO_2(g)$
$\text{sodium hydrogen carbonate}\quad\text{sodium carbonate}$

$\qquad 2Pb(NO_3)_2(s) \xrightarrow{\triangle} 2PbO(s) + 4NO_2(g) + O_2(g)$
$\text{lead (II) nitrate}\qquad\text{lead (II) oxide}\quad\text{nitrogen dioxide}$

e.g. 3 $\quad Fe(s) + H_2SO_4(aq) \longrightarrow H_2(g) + FeSO_4(aq)$
$\qquad\qquad\text{sulfuric acid}\qquad\qquad\text{iron (II) sulfate}$

$\qquad 2Li(s) + 2H_2O(l) \longrightarrow H_2(g) + 2LiOH(aq)$
$\qquad\qquad\qquad\qquad\text{lithium hydroxide}$
$\qquad\qquad\qquad\qquad\text{(in aqueous solution)}$

$\qquad Mg(s) + H_2O(g) \xrightarrow{\text{heating}} H_2(g) + MgO(s)$
$\qquad\qquad\text{(steam)}\qquad\qquad\text{magnesium oxide}$

$\qquad Sn(s) + 2HBr(aq) \longrightarrow H_2(g) + SnBr_2(aq)$
$\qquad\text{hydrobromic acid}\qquad\text{tin (II) bromide}$

$\qquad Zn(s) + CuCl_2(aq) \longrightarrow ZnCl_2(aq) + Cu(s)$
$\qquad\text{copper (II) chloride}\quad\text{zinc chloride}$

Activity series of metals

metal+cold $H_2O \longrightarrow H_2$ metal+steam $\longrightarrow H_2$ metal+acid $\longrightarrow H_2$
Li K Ba Sr Ca Na | Mg Al Mn Zn Cr | Fe | Ni Sn Pb | H_2 Sb Cu | Hg | Ag Pd Pt Au
--- | -------------------------------- | ---------------------- |
oxide+$H_2 \longrightarrow$ no reaction   oxide+$H_2 \longrightarrow$ metal   oxide+heat $\longrightarrow$ metal
---|
metal+$O_2 \longrightarrow$ oxide

e.g. 4 $MgCl_2(aq) + Ca(OH)_2(s) \longrightarrow Mg(OH)_2(s) + CaCl_2(aq)$
magnesium chloride   calcium hydroxide   magnesium hydroxide   calcium chloride
$BaCl_2(aq) + K_2CrO_4(aq) \longrightarrow BaCrO_4(s) + 2KCl(aq)$
barium chloride   potassium chromate   barium chromate   potassium chloride

4. Net Ionic Equation: Precipitation Reactions

Ions that are present during a reaction in aqueous solution but undergo no chemical change are called spectator ions. A net ionic equation shows only the species involved in a chemical change, excluding spectator ions. In such an equation charge must always be conserved; the sum of the charges on the left must equal the sum of the charges on the right. Whether a precipitate will form can be predicted from data on the solubility of different types of compounds, summarized below.

Generally Soluble
All Na^+, K^+, and NH_4^+ compounds
All Cl^-, Br^-, and I^- compounds
Except those of Ag^+, Pb^{2+}, Hg_2^{2+}, insol.
$PbCl_2$, sol. in hot water
$HgBr_2$, mod. sol.
I^- with heavier metals, insol.
All SO_4^{2-} compounds
Except those of Sr^{2+}, Ba^{2+}, Pb^{2+}, insol.
$CaSO_4$, $AgSO_4$, mod. sol.
All NO_3^- and NO_2^- compounds
Except $AgNO_2$, mod. sol.
All ClO_3^-, ClO_4^-, MnO_4^- compounds
Except $KClO_4$, mod. sol.
All CH_3COO^- compounds
Except $AgCH_3COO$, mod. sol.
Generally Insoluble
All S^{2-} compounds
Except those of NH_4^+, Li^+, Na^+, K^+, sol.
All O^{2-}, OH^- compounds
Except those of Li^+, Na^+, K^+, sol.
BaO, $Ba(OH)_2$, CaO, $Ca(OH)_2$, SrO, $Sr(OH)_2$, mod. sol.
All CO_3^{2-}, PO_4^{3-}, CN^-, SO_3^{2-} compounds
Except those of NH_4^+, Li^+, Na^+, K^+

第 5 章 化学反应和化学计算法 (Chemical Reactions and Stoichiometry)

e. g. 1
$$BaCl_2(aq) + K_2CrO_4(aq) \longrightarrow BaCrO_4(s) + 2KCl(aq)$$
$$\Downarrow$$
$$Ba^{2+}(aq) + 2Cl^-(aq) + 2K^+(aq) + CrO_4^{2-}(aq) \longrightarrow BaCrO_4(s) + 2K^+(aq) + 2Cl^-(aq)$$
$$\Downarrow$$
$$Ba^{2+}(aq) + CrO_4^{2-}(aq) \longrightarrow BaCrO_4(s)$$
(net ionic equation)

e. g. 2
$$MnS(s) + 2HCl(aq) \longrightarrow MnCl_2(aq) + H_2S(g)$$
$$\Downarrow$$
$$MnS(s) + 2H^+(aq) \longrightarrow Mn^{2+}(aq) + H_2S(g)$$

e. g. 3
$$MnO_4^-(aq) + 5Fe^{2+}(aq) + 8H^+(aq) \longrightarrow 5Fe^{3+}(aq) + Mn^{2+}(aq) + 4H_2O(l)$$
total charge: +17   total charge: +17

5.2 Stoichiometry

When chemical species (atoms, molecules, and ions) go into "action" we have chemical reactions—processes of chemical change. The calculation of the quantitative relationships in chemical change is called stoichiometry.

1. Information from Chemical Equations and Using Mole Ratios

All stoichiometric calculations must begin with the balanced chemical equation. The coefficients of the equation give the mole ratios of the various species. From these ratios, stoichiometry can give us information about the mass ratios of products and reactants, as well as volume ratios of any gases involved in the reaction.

e. g. 1
$$2SO_2(g) + O_2(g) \xrightarrow{catalyst} 2SO_3(g)$$

	sulfur dioxide		sulfur trioxide
molecular masses	64 u	32 u	80. u
	2 molecules	1 molecule	2 molecules
	2 moles	1 mole	2 moles
	2 volumes	1 volume	2 volumes
	128 u	32 u	160. u
	128 g	32 g	160. g
	44.8 L (STP)	22.4 L (STP)	44.8 L (STP)

2. Solving Stoichiometry Problems, Reactions Involving Gases, and Reactions in Aqueous Solution

In all stoichiometric problems, the mole ratios from the balanced chemical equation provide the connection between the known and unknown quantities, whether these be masses, pressure-volume-temperature data for gases, or molarities for substances in solution. To solve a stoichiometry problem, first write the balanced chemical equation, convert the known information to moles, use mole ratios to find the unknown in terms of moles, and convert the answer from moles to the desired quantity.

e. g. 1 How many grams of cobalt (II) chloride and of hydrogen fluoride are needed to prepare 10.0 g of cobalt (II) fluoride by the following reaction?

$$CoCl_2(s) + 2HF(g) \longrightarrow CoF_2(s) + 2HCl(g)$$

$(10.0 \text{ g CoF}_2) \times (1 \text{ mol CoF}_2/96.9 \text{ g CoF}_2) \times (1 \text{ mol CoCl}_2/1 \text{ mol CoF}_2) \times$
$(129.8 \text{ g CoCl}_2/1 \text{ mol CoCl}_2) = 13.4 \text{ g CoCl}_2$
$(10.0 \text{ g CoF}_2) \times (1 \text{ mol CoF}_2/96.9 \text{ g CoF}_2) \times (2 \text{ mol HF}/1 \text{ mol CoF}_2) \times$
$(20.0 \text{ g HF}/1 \text{ mol HF}) = 4.13 \text{ g HF}$

e.g. 2 What volume of H_2 (measured at 0.975 atm and 25°C) is required to react quantitatively with 46.6 g of WO_3 to displace tungsten?

$$3H_2(g) + WO_3(s) \xrightarrow{\text{heating}} 3H_2O(g) + W(s)$$

$(46.4 \text{ g WO}_3) \times (1 \text{ mol WO}_3/231.85 \text{ g WO}_3) \times (3 \text{ mol H}_2/1 \text{ mol WO}_3) = 0.600 \text{ mol H}_2$

$$V = \frac{nRT}{p} = \frac{(0.600 \text{ mol}) \times [0.0821 \text{ L} \cdot \text{atm}/(K \cdot \text{mol})] \times (298 \text{ K})}{0.975 \text{ atm}} = 15.1 \text{ L}$$

e.g. 3 How many milliliters of 0.10 mol/L barium chloride must be added to 25 mL of 0.23 mol/L solution of sodium sulfate to completely precipitate barium sulfate?

$$BaCl_2(aq) + Na_2SO_4(aq) \longrightarrow BaSO_4(s) + 2NaCl(aq)$$

$(25 \text{ mL}) \times (1 \text{ L}/1000 \text{ mL}) \times (0.23 \text{ mol Na}_2SO_4/1 \text{ L})$
$= 5.8 \times 10^{-3} \text{ mol Na}_2SO_4$
$(5.8 \times 10^{-3} \text{ mol Na}_2SO_4) \times (1 \text{ mol BaCl}_2/1 \text{ mol Na}_2SO_4)$
$= 5.8 \times 10^{-3} \text{ mol BaCl}_2$
$(5.8 \times 10^{-3} \text{ mol BaCl}_2) \times (1 \text{ L}/0.10 \text{ mol BaCl}_2) \times (1000 \text{ mL}/1 \text{ L})$
$= 58 \text{ mL}$

3. Limiting Reactants

The exact amount of a substance required by a balanced chemical equation is the stoichiometric amount. When reactants are present in non-stoichiometric amounts, the one that determines the amount of product that can be formed is called the limiting reactant. At the completion of the reaction, some of the other reactant(s) will be left over.

e.g. 1 What volume of hydrogen will be produced at 0.861 atm and 22°C from the displacement reaction of 6.0 g of zinc with 25 mL of 6.0 mol/L hydrochloric acid?

$$Zn(s) + 2HCl(aq) \longrightarrow ZnCl_2(aq) + H_2(g)$$

$(6.0 \text{ g Zn}) \times (1 \text{ mol Zn}/65.38 \text{ g Zn}) = 0.092 \text{ mol Zn}$
$(25 \text{ mL}) \times (1 \text{ L}/1000 \text{ mL}) \times (6.0 \text{ mol HCl}/1 \text{ L}) = 0.15 \text{ mol HCl}$
$(0.092 \text{ mol Zn}) \times (2 \text{ mol HCl}/1 \text{ mol Zn}) = 0.18 \text{ mol HCl}$
(The limiting reactant is HCl)
$(0.15 \text{ mol HCl}) \times (1 \text{ mol H}_2/2 \text{ mol HCl}) = 0.075 \text{ mol H}_2$

$$V = \frac{nRT}{p} = \frac{(0.075 \text{ mol})[0.0821 \text{ L} \cdot \text{atm}/(K \cdot \text{mol})](295 \text{ K})}{0.861 \text{ atm}} = 2.1 \text{ L}$$

4. Yields

The theoretical yield of a reaction is the maximum amount of a product that can be formed according to the balanced chemical equation. The actual yield may be less for various reasons. The percent yield represents the ratio of the actual yield to the theoretical yield expressed as a percentage.

e.g. 1 A chemist ran the following reaction

$$HOC_6H_4COOH \text{ (s)} + (CH_3CO)_2O \text{ (l)} \longrightarrow CH_3COOC_6H_4COOH \text{ (s)} + CH_3COOH \text{ (l)}$$
salicylic acid   acetic anhydride   acetylsalicylic acid (aspirin)   acetic acid

on a laboratory scale with 25.0 g of salicylic acid and excess acetic anhydride (over 20 g). The actual yield was 24.3 g of aspirin. What was the percent yield?

$$(25.0 \text{ g sal. acid}) \times (1 \text{ mol sal. acid}/138.13 \text{ g sal. acid}) \times$$
$$(1 \text{ mol aspirin}/1 \text{ mol sal. acid}) \times (180.17 \text{ g aspirin}/1 \text{ mol aspirin})$$
$$= 32.6 \text{ g aspirin}$$

$$\% \text{ yield} = [(\text{actual yield})/(\text{theoretical yield})](100\%)$$
$$= (24.3 \text{ g}/32.6 \text{ g}) \times (100\%) = 74.5\%$$

5. Stoichiometry in Industrial Chemistry

Calculations can be simplified by using such units as pound-moles or ton-moles instead of moles, in order to find masses in pounds, or tons, and so on, instead of in grams. For processes that include consecutive reactions, stoichiometric calculations can be based on overall equations and intermediate products can be disregarded if the intermediates cancel out of the equations. Also, partial equations can be used if the equations are balanced for the element or elements in question.

e.g. 1 Each day an electric power plant burns 4000 t of coal which contains 1.2% S by mass. During the combustion process, the sulfur is quantitatively converted to sulfur dioxide. Calculate the mass (in grams) of sulfur dioxide produced each day by the plant. (1 t = 9.07×10^5 g.)

$$S \text{ (s)} + O_2 \text{ (g)} \longrightarrow SO_2 \text{ (g)}$$
$$(4000 \text{ t coal}) \times (1.2 \text{ t S}/100.0 \text{ t coal}) = 50 \text{ t S}$$
$$(50 \text{ t S}) \times (1 \text{ t-mol S}/32.06 \text{ t S}) = 2 \text{ t-mol S}$$
$$(2 \text{ t-mol S}) \times (1 \text{ t-mol } SO_2/1 \text{ t-mol S}) \times (64.06 \text{ t } SO_2/1 \text{ t-mol } SO_2)$$
$$= 100 \text{ t } SO_2 = 1 \times 10^8 \text{ g } SO_2$$

e.g. 2 Adding the following two equations allows cancellation of the intermediate, potassium manganate.

$$2MnO_2 \text{ (s)} + 4KOH \text{ (aq)} + O_2 \text{ (g)} \longrightarrow 2K_2MnO_4 \text{ (aq)} + 2H_2O \text{ (l)}$$
$$2K_2MnO_4 \text{ (aq)} + Cl_2 \text{ (g)} \longrightarrow 2KMnO_4 \text{ (aq)} + 2KCl \text{ (aq)}$$

e.g. 3 Calculate the mass of $Na_5P_3O_{10}$ that can be produced from the complete conversion of phosphorous obtained from processing 60. t of the ore which contains 45% $Ca_3(PO_4)_2$.

$$3Ca_3(PO_4)_2 + \ldots \longrightarrow 2Na_5P_3O_{10} + \ldots$$
$$(60. \text{ t ore}) \times (45 \text{ t } Ca_3(PO_4)_2/100. \text{ t ore}) = 27 \text{ t } Ca_3(PO_4)_2$$
$$(27 \text{ t } Ca_3(PO_4)_2) \times (1 \text{ t-mol } Ca_3(PO_4)_2/310.18 \text{ t } Ca_3(PO_4)_2)$$
$$(2 \text{ t-mol } Na_5P_3O_{10}/3 \text{ t-mol } Ca_3(PO_4)_2) \times (367.86 \text{ t } Na_5P_3O_{10}/1 \text{ t-mol } Na_5P_3O_{10})$$
$$= 21 \text{ t } Na_5P_3O_{10}$$

Homework No. 5

01 Balance and classify each of the following chemical equations as a (i) combination reaction, (ii) decomposition reaction, (iii) displacement reaction, or (iv) partner-exchange reaction:

(a) $Fe_3O_4(s) + H_2(g) \longrightarrow Fe(s) + H_2O(l)$

(b) $KClO_3(s) \longrightarrow KCl(s) + O_2(g)$

(c) Steam and hot carbon react to form gaseous hydrogen and gaseous carbon monoxide

(d) $Cl_2O_7(g) + H_2O(l) \longrightarrow HClO_4(aq)$

(e) $Br_2(l) + H_2O(l) \longrightarrow HBr(aq) + HBrO(aq)$

(f) $Ca_3(PO_4)_2(s) + H_2SO_4(aq) \longrightarrow CaSO_4(s) + H_3PO_4(aq)$

(g) Potassium reacts with water to give aqueous potassium hydroxide and gaseous hydrogen

(h) Solid magnesium carbonate decomposes to form solid magnesium oxide and gaseous carbon dioxide.

02 Use the activity series to predict whether or not the following reactions will occur:

(a) $Fe(s) + Mg^{2+} \longrightarrow Mg(s) + Fe^{2+}$

(b) $Ni(s) + Cu^{2+} \longrightarrow Ni^{2+} + Cu(s)$

(c) $Cu(s) + 2H^+ \longrightarrow Cu^{2+} + H_2(g)$

(d) $Mg(s) + H_2O(g) \longrightarrow MgO(s) + H_2(g)$

(e) $Sn(s) + Ba^{2+} \longrightarrow Sn^{2+} + Ba(s)$

(f) $Al_2O_3(s) + 3H_2(g) \xrightarrow{\triangle} 2Al(s) + 3H_2O(g)$

(g) $Ca(s) + 2H^+ \longrightarrow Ca^{2+} + H_2(g)$

(h) $Cu(s) + Pb^{2+} \longrightarrow Cu^{2+} + Pb(s)$

03 Based on the solubility rules given in the lecture, how would you write the formulas for the following substances in a net ionic equation: (a) $PbSO_4$, (b) $Na(CH_3COO)$, (c) $(NH_4)_2CO_3$, (d) MnS, (e) $BaCl_2$, (f) $(NH_4)_2SO_4$, (g) $NaBr$, (h) $Ba(CN)_2$, (i) $Mg(OH)_2$, and (j) Li_2CO_3.

04 Consider the hypothetical equation describing photosynthesis:

$$CO_2(g) + H_2O(l) \xrightarrow[\text{chlorophyll}]{\text{light}} C_6H_{12}O_6(aq) + O_2(g) \quad \text{(unbalanced)}$$

(a) Identify the (i) reactants, (ii) products, and (iii) catalyst.

(b) What else is needed for the reaction to occur?

(c) Balance the equation.

(d) What is the mole ratio of (i) O_2 to CO_2, (ii) $C_6H_{12}O_6$ to O_2, and (iii) CO_2 to H_2O?

(e) Does the reaction obey the law of conservation of mass?

(f) If the reaction were carried out in a closed container, would there be a pressure change?

05 Nitrous oxide, N_2O, undergoes decomposition when heated to give N_2 and O_2.

$$2N_2O(g) \xrightarrow{\triangle} 2N_2(g) + O_2(g)$$

What is the molar composition of the gaseous mixture produced? Compare this composition to that of air and predict whether the mixture will support combustion or not.

06 A sheet of iron was galvanized (plated with zinc) on both sides to protect it from rust. The thickness of the zinc coating was determined by allowing hydrochloric acid to react with the zinc and collecting the resulting hydrogen. (Note: The acid solution contained an "inhibitor" ($SbCl_3$) which prevented the iron from reacting.)

$$Zn\ (s) + 2HCl\ (aq) \longrightarrow ZnCl_2\ (aq) + H_2\ (g)$$

Determine the thickness of the zinc plate from the following data: sample size = 1.50 cm × 2.00 cm; volume of dry hydrogen = 30.0 mL; temperature = 25°C; pressure = 747 Torr; and density of zinc = 7.11 g/cm³.

07 Consider the following reaction:

$$HNO_3(aq) + Cu\ (s) \longrightarrow Cu(NO_3)_2(aq) + NO_2(g) + H_2O\ (l)$$

(a) Balance the equation. A piece of Cu metal 3.31 cm × 1.84 cm × 1.00 cm reacts with 157 mL of 1.35 mol/L nitric acid solution. The density of copper is 8.92 g/cm³.

(b) Find the number of moles of each reactant.

(c) What volume of NO_2 at 1.01 atm and 297 K will be formed?

(d) Describe what would happen if the amount of Cu were doubled.

08 Hydrogen reacts with some of the more active metals to form crystalline ionic hydrides. For example, Li forms LiH

$$2Li\ (s) + H_2(g) \longrightarrow 2LiH\ (s)$$

(a) What mass of LiH would be produced by allowing 10.0 g of Li to react with 10.0 L of H_2 (measured at STP)?

(b) If the actual yield was 6.7 g of LiH, what is the percent yield?

09 What mass of H_2SO_4 can be produced in the process given below if 1.00 kg of FeS_2 is used? The unbalanced equations for the process are

$$FeS_2(s) + O_2(g) \longrightarrow Fe_2O_3(s) + SO_2(g)$$
$$SO_2(g) + O_2(g) \longrightarrow SO_3(g)$$
$$SO_3(g) + H_2SO_4(l) \longrightarrow H_2S_2O_7(l)$$
$$H_2S_2O_7(l) + H_2O\ (l) \longrightarrow H_2SO_4(aq)$$

第6章 热化学
Chapter 6　Thermochemistry

6.1　Energy

1. Energy in Chemical Reactions

All chemical reactions are accompanied by energy changes. In general, breaking chemical bonds requires energy and the formation of chemical bonds releases energy.

2. Thermodynamics

In thermodynamics—the study of energy transformations—we define the area of study as the system and consider the rest of the universe to be the surroundings. (A system in chemistry is usually the substances undergoing a physical or chemical change.) Heat is the energy transferred between objects that differ in temperature; it represents the random kinetic energy (energy of motion) of atoms and molecules. Any process that releases heat from a system to its surroundings is exothermic; any process in which a system absorbs heat from its surroundings is endothermic.

3. Internal Energy

All of the energy contained by a chemical system is classified as internal energy. An increase in internal energy can have three consequences: It can raise the temperature, it can cause a phase change, or (if it is sufficient to break chemical bonds, allowing new ones to form) it can produce a chemical reaction. The internal energy change of a system is symbolized by ΔE.

4. Energy, Heat, and Work

The first law of thermodynamics, known as the law of conservation of energy, states that the energy of the universe is constant. Thus any energy lost by a system must be transferred to its surroundings, and vice versa. The entire energy change must be accounted for by heat and/or work. Mathematically, $\Delta E = Q + W$. Work is preformed when a force moves an object over a distance. For a gas expanding against constant pressure, the work, W, is equal to $-p\Delta V$. At constant volume, the flow of heat in a chemical change is equal to the change in internal energy: $\Delta E = Q_v$. Enthalpy, H, is a thermodynamic property of a system defined by the formula $H = E + pV$. The change in enthalpy in a process at constant pressure is equal to the amount of heat exchanged with the surroundings: $\Delta H = Q_p$. H has units of energy.

e. g. 1 In a single process, a system does 125 J of work on its surroundings while 75 J of heat is added to the system. What is the internal energy change for the system?
$W = -125$ J, $Q = 75$ J,
$\Delta E = Q + W = 75$ J $+ (-125$ J$) = -50$ J.

e. g. 2 For each of the following chemical and physical changes at constant pressure, is work done by the system (the substances undergoing the change) on the surroundings, or by the surroundings on the system, or is the amount of work negligible?

(a) $Sn(s) + 2F_2(g) \longrightarrow SnF_4(s)$
  $\Delta V < 0, W > 0$

(b) $AgNO_3(aq) + NaCl(aq) \longrightarrow AgCl(s) + NaNO_3(aq)$
  negligible (no gases)

(c) $C(s) + O_2(g) \longrightarrow CO_2(g)$
  $\Delta V = 0$, negligible

(d) $SiI_4(g) \xrightarrow{\Delta} Si(s) + 2 I_2(g)$
  $\Delta V > 0, W < 0$

6.2 Heats of Reaction and Other Enthalpy Changes

1. Heats of Reaction

Many chemical reactions are carried out at constant pressure (usually atmospheric). Under these conditions, the heat of reaction—the amount of heat released or absorbed from the start of the process to the time when the system has returned to its original temperature—is equal to the enthalpy change, ΔH. A thermochemical equation is one that includes ΔH for the balanced equation as written. For exothermic processes, ΔH is negative; for endothermic processes, ΔH is positive.

e. g. 1 $2HgO(s) \xrightarrow{\Delta} 2Hg(l) + O_2(g) \quad \Delta H_{298} = 181.67$ kJ

$HgO(s) \xrightarrow{\Delta} Hg(l) + (1/2)O_2(g) \quad \Delta H_{298} = 90.84$ kJ

$2HgO(s) + 181.67$ kJ $\longrightarrow 2Hg(l) + O_2(g)$

2. Standard State and Standard Enthalpy Changes, Heats of Formation, and Heats of Combustion

The standard state of any susbstance is the physical state in which it is most stable at 1 atm and a specified temperature (usually 298 K). Enthalpy changes for chemical reactions of substances in their standard states are known as standard enthalpy changes ($\Delta H^\ominus$). The standard enthalpy of formation of a compound, $\Delta H_f^\ominus$ is the heat of formation of one mole by combination of its elements in their standard states at a specified temperature. The standard enthalpy of combustion, $\Delta H_c^\ominus$ is the heat for the reaction of one mole of a substance in its standard state with oxygen.

e. g. 1 In their standard states at 25°C, calcium is a solid (mp$>$25°C), $CH_3(CH_2)_6CH_3$ is a liquid (mp$<$25°C, bp$>$25°C), and GeH_4 is a gas (bp$<$25°C).

e.g. 2 $H_2(g) + (1/2)O_2(g) \longrightarrow H_2O(l)$ $\Delta H_f^\ominus = -285.83 \text{ kJ}$

e.g. 3 $CH_4(g) + 2O_2(g) \longrightarrow CO_2 + 2H_2O(l)$ $\Delta H_c^\ominus = -890. \text{ kJ}$

3. Finding Enthalpy Changes and Heats of Reaction from Standard Heats of Formation

The enthalpy change, ΔH, for a reaction is equal in magnitude to ΔH for the reverse reaction but opposite in sign. If the quantities of reactants and products in a reaction are changed, ΔH for that reaction changes proportionately. For any chemical reaction, ΔH has the same value whether the reaction takes place in one step or in several steps (Hess's law). This means that thermochemical equations can be manipulated and combined algebraically to find unknown enthalpy changes from known ones (e.g., $\Delta H^\ominus = \Delta H_f^\ominus$ of products $- \Delta H_f^\ominus$ of reactants).

e.g. 1 $Ag_2O(s) \xrightarrow{\triangle} 2Ag(s) + (1/2)O_2(g)$ $\Delta H^\ominus = -\Delta H_f^\ominus = 31.05 \text{ kJ}$

e.g. 2 $2Ag_2O(s) \xrightarrow{\triangle} 4Ag(s) + O_2(g)$ $\Delta H^\ominus = 2 \times 31.05 \text{ kJ} = 62.10 \text{ kJ}$

e.g. 3 How much heat will be released in the following reaction:

$$2Al(s) + Cr_2O_3(s) \longrightarrow Al_2O_3(s) + 2Cr(s) \quad \Delta H^\ominus = -536 \text{ kJ}$$

(under standard state conditions at 298 K) of 10.0 g of aluminum with 25.0 g of Cr_2O_3?

$$(10.0 \text{ g Al}) \times (1 \text{ mol Al} / 26.98 \text{ g Al}) = 0.371 \text{ mol Al}$$

$$(25.0 \text{ g } Cr_2O_3) \times (1 \text{ mol } Cr_2O_3 / 151.99 \text{ g } Cr_2O_3) = 0.164 \text{ mol } Cr_2O_3$$

$$(0.371 \text{ mol Al}) \times (1 \text{ mol } Cr_2O_3 / 2 \text{ mol Al}) = 0.186 \text{ mol } Cr_2O_3$$

Cr_2O_3 is the limiting reactant.

$$(0.164 \text{ mol } Cr_2O_3) \times (-536 \text{ kJ} / 1 \text{ mol } Cr_2O_3) = -87.9 \text{ kJ}$$

e.g. 4 Combine the following thermochemical equations

$$N_2O_4(g) \longrightarrow 2NO_2(g) \quad \Delta H^\ominus = 57.20 \text{ kJ}$$

$$2NO(g) + O_2(g) \longrightarrow 2NO_2(g) \quad \Delta H^\ominus = -114.14 \text{ kJ}$$

to find the heat of reaction for

$$2NO(g) + O_2(g) \longrightarrow N_2O_4(g).$$

$\underline{2NO_2(g)} \longrightarrow N_2O_4(g)$ $\Delta H^\ominus = -57.20 \text{ kJ}$

$2NO(g) + O_2(g) \longrightarrow \underline{2NO_2(g)}$ $\Delta H^\ominus = -114.14 \text{ kJ}$

$2NO(g) + O_2(g) \longrightarrow N_2O_4(g)$ $\Delta H^\ominus = -171.34 \text{ kJ}$

e.g. 5 Find the heat of reaction for

$$CH_4(g) + 4F_2(g) \longrightarrow CF_4(g) + 4HF(g)$$

using the following heat-of-formation data:

$\Delta H_f^\ominus (CF_4) = -925 \text{ kJ/mol}$

$\Delta H_f^\ominus (HF) = -271.1 \text{ kJ/mol}$

$\Delta H_f^\ominus (CH_4) = -74.81 \text{ kJ/mol}$

$\Delta H_f^\ominus (F_2) = 0 \text{ kJ/mol}$

$\Delta H^\ominus = (1 \text{ mol}) \times (-925 \text{ kJ/mol}) + (4 \text{ mol}) \times (-271.1 \text{ kJ/mol}) - (1 \text{ mol}) \times (-74.81 \text{ kJ/mol}) - (4 \text{ mol}) \times (0 \text{ kJ/mol}) = -1935 \text{ kJ}$

4. Heats of Changes of State

The enthalpy for a change of state is the amount of heat absorbed or released at constant pressure and without any change in temperature, for example, in melting or freezing, evaporation or condensation.

$$\text{solid} \underset{\text{crystallization}}{\overset{\text{fusion}}{\rightleftharpoons}} \text{liquid} \underset{\text{condensation}}{\overset{\text{vaporization}}{\rightleftharpoons}} \text{gas}$$

$$\text{solid} \underset{\text{deposition}}{\overset{\text{sublimation}}{\rightleftharpoons}} \text{gas}$$

6.3 Measuring Heat

1. Heat Capacity

The heat capacity of a substance is the amount of heat needed to raise the temperature of a given amount of the substance by 1 K. The amount of the substance may be 1 mol (molar heat capacity, C_p) or 1 g (specific heat).

e. g. 1 How much is needed to raise the temperature of 21 g of aluminium from 25°C to 161°C (no phase changes occur)? The specific heat of aluminum is 0.902 J/(K·g).

$$\Delta T = (273+161) \text{ K} - (273+25) \text{ K} = 136 \text{ K}$$

$$Q = (\text{mass}) \times (\text{specific heat}) \Delta T = (21 \text{ g}) \times [0.902 \text{ J}/(\text{K}\cdot\text{g})] \times (136 \text{ K}) = 2600 \text{ J}$$

2. Calorimetry

A calorimeter is used for measuring the heat flow during a thermochemical process. The heat of the process is found from the change in temperature of the calorimeter and its contents, ΔT, and the heat capacity of the calorimeter (which must be experimentally determined).

e. g. 1 A calorimeter with a C_p of 60 J/K was used to find the heat flow associated with the dissolution of 24 g of NaCl in 176 g of water. The observed ΔT was -1.6 K. The specific heat of the resulting NaCl solution is 3.64 J/(K·g). What is the heat flow for this process?

$$Q_{\text{dissoln}} + Q_{\text{calorimeter}} + Q_{\text{NaCl soln}} = 0$$

$$Q_{\text{lost, NaCl soln}} = (176 \text{ g} + 24 \text{ g}) \times [3.64 \text{ J}/(\text{K}\cdot\text{g})] \times (-1.6 \text{ K}) = -1200 \text{ J}$$

$$Q_{\text{lost, calorimeter}} = (60 \text{ J/K}) \times (-1.6 \text{ K}) = -100 \text{ J}$$

$$Q_{\text{dissoln}} = -[-1200 \text{ J} + (-100 \text{ J})] = 1300 \text{ J}$$

Homework No. 6

01 For each of the following chemical and physical changes carried out at constant pressure, decide whether work is done by the system (the substances undergoing the change) on the surroundings or by the surroundings on the system or whether the amount of work is negligible.

(a) $C_6H_6(l) \longrightarrow C_6H_6(s)$

(b) $(1/2)N_2(g) + (3/2)H_2(g) \longrightarrow NH_3(g)$

(c) $3H_2S(g) + 2HNO_3(g) \longrightarrow 2NO(g) + 4H_2O(l) + 3S(s)$

02 Suppose it were possible to melt a substance and then vaporize it at the same temperature. What would be the relationship among the heats of fusion, vaporization, and sublimation?

03 The following processes were studied at 25°C.

(a) $CaSO_4 \cdot 5H_2O\,(s) \longrightarrow CaSO_4\,(s) + 5H_2O\,(g)$

(b) $2NH_3\,(g) + H_2SO_4\,(aq) \longrightarrow (NH_4)_2SO_4\,(aq)$

Assuming molar quantities of reactants and products, the work that the system exchanges with its surroundings in each case can be calculated using $W = -RT$ (sum of n of gaseous products — sum of n of gaseous reactants). Calculate the work for each of the reactions assuming a constant external pressure of 1.00 atm.

04 The heat of formation at 1000 K for Al (s) is -10.519 kJ/mol, for Al (l) it is 0.000 kJ/mol, and for Al (g) it is 310.114 kJ/mol. Find the heat of (a) fusion, (b) vaporization, and (c) sublimation at this temperature.

05 A sample of coke contains 90.9% carbon by mass. Assuming that the heat produced by the burning of this coke comes from the combustion of C to CO_2, calculate the total quantity of heat obtainable at 25°C through the burning of exactly 1 kg of the coke. If the coke sample contained 0.1% sulfur by mass and this sulfur burned completely to sulfur dioxide, SO_2, what is the total quantity of heat that would result from this source when the coke was burned? The heats of combustion of C and S are -394 kJ/mol and -297 kJ/mol, respectively.

06 In 1819 Pierre Dulong and Alexis Petit recognized that for all solid metallic elements at room temperature (except for a few with very small atomic masses) the product of the atomic mass of the element and its specific heat is approximately a constant

$$\text{(atomic mass)} \times \text{(specific heat)} = 26 \text{ J/(K} \cdot \text{mol)}$$

Using the following specific heat data, show that this empirical relationship is true.

Al	0.900 J/(K·g)	Fe	0.452 J/(K·g)
Be	1.824	Pb	0.130
Cr	0.460	Sn	0.226

07 A temperature increase of 54.0°C was observed upon the addition of 76.5 J of heat to a 2.71 g sample of an unknown metal. Calculate the specific heat of the metal. Using the rule of Dulong-Petit (see Problem 06), identify the unknown metal.

08 (a) A student heated a sample of a metal weighing 32.6 g to 99.83°C and put it into 100.0 g of water at 23.62°C in a calorimeter. The final temperature was 24.41°C. The student calculated the specific heat of the metal, neglecting to use the heat capacity of the calorimeter. The specific heat is 4.184 J/(K·g) for H_2O. What was his answer?

The metal was known to be either Cr, Mo, or W and by comparing the value of the specific heat to those of the metals [Cr, 0.460; Mo, 0.250; W, 0.135 J/(K·g)], the student identified the metal. What was the metal?

(b) A student at the next laboratory bench did the same experiment, got the same data, and used the heat capacity of the calorimeter in his calculations. The heat capacity of the calorimeter was 410 J/K. Was his identification of the metal different?

09 The heat of formation of HCl (g) is -92.307 kJ/mol at 25°C. The value of $\Delta H_f^{\ominus}$ at 500 K can be found by:

(a) calculating the enthalpy change for cooling 1/2 mol of $H_2\,(g)$ and 1/2 mol of $Cl_2\,(g)$ from 500 K to 298 K.

(b) adding the heat of reaction at 298 K to the answer to part (a).

(c) adding the enthalpy change for heating one mole of HCl (g) from 298 K to 500 K to the result of part (b).

$$\frac{1}{2}H_2(g) + \frac{1}{2}Cl_2(g) \xrightarrow{\Delta H^{\ominus}_{500} = ?} HCl(g)$$
$$\downarrow \Delta H^{\ominus}_a \qquad\qquad\qquad \uparrow \Delta H^{\ominus}_c$$
$$\frac{1}{2}H_2(g) + \frac{1}{2}Cl_2(g) \xrightarrow{\Delta H^{\ominus}_{298,b}} HCl(g)$$

Find $\Delta H^{\ominus}_f$ at 500 K, given the heat capacities of 33.907 J/(K·mol) for $Cl_2(g)$, 29.12 J/(K·mol) for HCl (g), and 28.824 J/(K·mol) for $H_2(g)$. Is the reaction more or less exothermic at 500 K than at 298 K?

第7章 有机化合物和基团的命名
Chapter 7 Nomenclature for Organic Compounds and Groups

7.1 Saturated and Unsaturated Hydrocarbons

1. Nomenclature for Saturated Hydrocarbons

(i) Alk**anes** (paraffin hydrocarbons) C_nH_{2n+2}

(a) The first four members of the alkane series

meth**ane** (CH_4), eth**ane** (CH_3CH_3), prop**ane** ($CH_3CH_2CH_3$), but**ane** ($CH_3CH_2CH_2CH_3$)

(b) Alkane with more than four carbon atoms

• A prefix (penta, C_5; hexa, C_6; hepta, C_7; octa, C_8; nona, C_9; deca, C_{10}) indicates the longest continuous chain of carbon atoms; the suffix **-ane** is added to this prefix (one of the two "a's" is dropped). e.g., pentane, hexane.

• The position and name of branches from the main chain, or of atoms other than hydrogen, are added as prefixes to the name of the longest hydrocarbon chain. The position of attachment to the longest continuous chain is given by a number obtained by numbering the longest chain from the end nearest the branch. In this way, the groups attached to the chain are designated by the lowest numbers. e.g., 2-chloropentane, 2-methylbutane, 2,2,4-trimethylpentane (or the common name, isooctane), 1-chloro-3,3-dimethylpentane, 2-methylpropane (isobutane).

(ii) **Cyclo**alkanes (alicyclic hydrocarbons or cycloparaffin) C_nH_{2n}

The nomenclature of cycloalkanes follows the same pattern used for the noncyclic alkanes. If there are substitutes, one is given the number 1 and others are given the lowest possible numbers. e.g., **cyclo**propane, 1,2-dichloro**cyclo**hexane.

(iii) Primary (the C joined to only one other carbon), secondary (to two other C's, as in the sec-butyl group), tertiary (as in the tert-butyl group), and quarternary carbon atom.

(iv) normal hydrocarbons (unbranched): n-hexane or hexane.

(v) methyl, ethyl, n-propyl, isopropyl, n-butyl, tert-butyl, n-pentyl, cyclopropyl, cyclobutyl, cyclohexyl.

2. Nomenclature for Unsaturated Hydrocarbons

(i) Alk**enes** (olefins) C_nH_{2n} and alkynes (acetylenes) C_nH_{2n-2}

To derive the systematic names of the individual alkenes, the **-ane** of the corresponding saturated hydrocarbon name is dropped and **-ene** is added if one double bond is present, **adi**ene is added if two double bonds are present, and so on. e. g., ethene, 1,3-butadiene.

Similarly, the individual alkynes are named by dropping the **-ane** and adding **-yne**, adiyne, atriyne, and so on. In either case, the position of the multiple bond is indicated by numbering from the end of the chain, starting at the end that will assign the lower number to the first carbon atom of the multiple bond. e. g., 1,3-pentadiyne.

In the common system of nomenclature, the **-ane** ending of the saturated hydrocarbon name is replaced by **-ylene** for the olefins. Compounds containing triple bonds are sometimes named as substituted acetylenes. (Because of the triple bond between the carbon atoms in acetylene, each carbon can have only one group attached to it.) e. g., ethylene (ethene, vinyl), propylene (propene, allyl), α-butylene (1-butene), isobutylene (2-methylpropene) β-butylene (2-butene, crotyl); acetylene (ethyne, ethynyl), methylacetylene (propyne, propargyl), vinylacetylene (1-buten-3-yne).

Cycloalkenes are common, but cycloalkynes exist only for C_8 or larger rings. The triple bond is not flexible enough to fit easily into smaller rings.

Beginning with propylene, alkenes of the homologous series of molecular formula C_nH_{2n} are isomeric with cycloalkanes. (e. g., propene and cyclopropane.) Acetylenes of the series of molecular formula C_nH_{2n-2} with $n>2$ are isomeric with cycloalkenes. (e. g., propyne and cyclo-propene.)

(ii) Aromatic hydrocarbons and polycyclic aromatic hydrocarbons

In aromatic hydrocarbons derived from benzene, the added groups are referred to as substituents. The name of a single substituent is added to "benzene" as a prefix, as in ethylbenzene. Three structurally isomeric forms are possible for a disubstituted benzene, whether or not the substituents are the same. The three possibilities are designated ortho (abbreviated *o-*), meta (abbreviated *m-*), and para (abbreviated *p-*), as may be seen in the xylenes, e. g., *m*-xylene ($C_6H_4(CH_3)_2$).

Numbers are also used to show the positions of substituents in aromatic compounds. Unless there is no question of what the structure is, as in hexachlorobenzene, numbers are always used to locate the substituents when three or more are present in the ring.

The benzene molecule less one hydrogen atom is known as the **phenyl group**, C_6H_5. Diphenylmethane, for example, is $(C_6H_5)_2CH_2$.

Polycyclic aromatic hydrocarbons contain two or more aromatic rings fused together. ("Fused" rings have in common a bond between the same two atoms.) Some examples are: naphthalene, 1,4-dimethylnaphthalene, anthracene, and phenanthrene. Several resonance forms can be written for each of these compounds. Numbers are assigned by convention to carbon atoms in fused ring systems (except to those at the points of fusion, where substitution is not possible). The location of substituents are identified by the assigned numbers.

naphthalene   anthracene

3. Nomenclature for Functional Groups

halo (—X), **hydroxyl** (—OH), **alkoxy** (—OR), **amino** [—NR$_3$; R$_n$NH$_{3-n}$, a primary ($n=1$) or secondary ($n=2$) or tertiary amine ($n=3$)], **aldehyde** or **formyl** (—CHO), **carbonyl** or **keto** or **oxo** (—CO—), **carboxyl** (—COOH), **amido** (—CONH$_2$), **carbonyl halide** (—COX), **anhydride** (—COOCOR), **ester** (—COOR), **nitro** (—NO$_2$; nitroalkane R—NO$_2$), **sulfonic acid** (—SO$_3$H; sulfonic acid R—SO$_3$H), **cyano** (—CN; nitrile R—CN).

7.2 Functional Groups with Covalent Single Bonds

Alkyl and **aryl halides** (RX, ArX), **alcohols** (ROH), **phenols** (ArOH), **ethers** (ROR, where the R's may be the same or different), and **amines** (primary, RNH, and secondary, R$_2$NH, or tertiary, R$_3$N, in which the R's may be the same or different) contain the following functional groups in which there are only single bonds: —X, —OH, —OR, and —NH$_2$, —NHR, —NR$_2$. Alcohols are classified as primary, secondary, or tertiary depending on whether the carbon atom to which the OH group is attached is primary, secondary, or tertiary. Alcohols exhibit hydrogen bonding and those with low molecular masses are miscible with water. Phenols are weak acids. Ethers tend to be un-reactive, and low molecular mass ethers are often used as solvents. The amines are organic bases.

1. Nomenclature for Functional Groups with Single Bonds

(i) alkyl or aryl halides (RX or ArX)

methyl bromide (bromomethane), methyl iodide (iodomethane), ethyl bromide (bromoethane), propyl bromide (1-bromopropane), propylene dibromide (1, 2-dibromopropane), vinyl chloride (chloroethene), chlorobenzene.

(ii) alcohols (ROH) and phenols (ArOH)

In the IUPAC system of nomenclature, the name is derived from the longest hydrocarbon chain that includes the OH group by dropping the final **-e** and adding **-ol**, as in methanol, ethanol, and cyclohexanol. When necessary, a number is used to show the position of the OH group. Numbering starts at the end of the chain nearest to the OH group. e.g., methyl alcohol (methanol), ethyl alcohol (ethanol), propyl alcohol (1-propanol), isopropyl alcohol (2-propanol), 2-butanol, cyclopentanol, tert-butyl alcohol, ethylene glycol (1,2-ethanediol), glycerol (glycerin, 1,2,3-propanetriol), β-phenylethyl alcohol (2-phenylethanol); alkoxide $RO^- M^+$, e. g., CH_3CH_2ONa, sodium ethoxide; phenol (C_6H_5OH), p-cresol ($CH_3C_6H_4OH$), α-naphthol (1-naphthol), catechol (o-$C_6H_4(OH)_2$), hydroquinone (p-$C_6H_4(OH)_2$).

(iii) Ethers (ROR')

dimethyl ether or methyl ether (methoxymethane), diethyl ether (ethoxyethane), ethylene glycol dimethyl ether ("glyme") (1, 2-dimethoxyethane), ethylene oxide (oxirane), dioxane (1,4-dioxin), anisole (methoxybenzene).

(iv) Amines (R_nN)

methylamine (CH_3NH_2), dimethylamine (($CH_3)_2NH$), trimethylamine (($CH_3)_3N$), tetramethylammonium iodide (($CH_3)_4NI$, a quaternary ammonium salt); aniline ($C_6H_5NH_2$), β-naphthylamine (2-naphthylamine); pyridine, morpholine.

7.3 Functional Groups with Covalent Double Bonds

Aldehydes (RCHO) and ketones (RCOR) (in which the R's may be the same or different) have carbonyl (—CO—)-containing functional groups. A secondary alcohol can be oxidized to a ketone. A primary alcohol can be oxidized to an aldehyde, and further oxidation yields a carboxylic acid (RCOOH) which contains the carboxyl group (—COOH). Carboxylic acids are generally weak acids and form salts when treated with bases. Esters (RCOOR), acyl halides (RCOX), acid anhydrides (RCOOCOR), and amides (RCONH₂ and the N-substituted amides RCONHR and RCONR₂) are common types of organic compounds which have carbonyl-containing functional groups that are derived from carboxyl groups.

1. Nomenclature for Functional Groups with Double Bonds

(i) aldehydes (RCHO) and ketones (RCOR')

formaldehyde (methanal), acetaldehyde (ethanal), caproaldehyde (hexanal), acrolein (propenal), benzaldehyde

dimethyl ketone (acetone, 2-propanone), biacetyl (dimethylglyoxal, 2,3-butanedione), methyl n-propyl ketone (2-pentanone), methyl phenyl ketone (acetophenone)

(ii) carboxylic acids (RCOOH)

formic acid (methanoic acid), acetic acid (ethanoic acid), propionic acid (propanoic acid), acrylic acid (propenoic acid), butyric acid (butanoic acid); benzoic acid, phthalic acid (o-$C_6H_4(COOH)_2$), terephthalic acid (p-$C_6H_4(COOH)_2$); lactic acid, tartaric acid, citric acid; sodium acetate (sodium ethanoate)

(iii) esters (RCOOR')

In the two-word name for an ester, the first word is the name of the R' group in RCOOR'. This word could be methyl, ethyl, phenyl, or the like. The second word of the name is the name of the carboxylic acid with the final -ic replaced by -ate, identical with the name of the carboxylic anion.

methyl formate (methyl methanoate), methyl acetate (methyl ethanoate), ethyl acetate (ethyl ethanoate), methyl acrylate (methyl propenoate); methyl benzoate ($C_6H_5COOCH_3$), diethyl phthalate (o-).

Other kinds of esters include lactaone, e. g., caprolactone (from caproic acid).

(iv) acyl halides (RCOX) and carboxylic acid anhydrides(RCOOCOR')

An acyl group (RCO—) is named by dropping the final -ic from the name of the corresponding carboxylic acid, and adding -yl. The complete name of the acyl halide is then the name of the acyl group, followed by the anionic name for the halogen atom, e. g., acetyl chloride (ethanoyl chloride), benzoyl chloride.

Anhydrides are given the name of the corresponding acid, followed by the word anhydride. Mixed anhydrides, from two different carboxylic acids, and cyclic anhydrides, are named similarly, for instance, acetic butyric anhydride, $(CH_3CO)O(COC_3H_7)$. e. g., acetyl chloride (ethanoyl chloride), benzoyl chloride, acetic anhydride (ethanoic anhydride), benzoic anhydride, phthalic anhydride.

(v) amides ($RCONH_2$)

The common name for an amide is derived by replacing the -yl of the acyl name by amide. Thus, acetamide is the name for CH_3CONH_2. Since the acyl group is named by replacing the -ic of the carboxylic acid name by -yl, one can equally well derive the amide name from the corresponding acid name by dropping the -ic (or -oic of the IUPAC name) and adding amide. If one or both of the hydrogen atoms on the amide nitrogen atom are replaced by hydrocarbon groups, the structure is named as an N-substituted amide.

acetamide (ethanamide), benzamide, N,N-dimethyl-formamide, N-phenylacetamide (acetanilide); sulfanilamide.

Other kinds of amides include lactam or lactan, e. g., caprolactam.

Homework No. 7

01 What is the name of the international body that devised the systematic nomenclature used for organic compounds? In this system what is the suffix that is used to indicate a saturated hydrocarbon? How is the number of carbon atoms indicated in the name of a saturated hydrocarbon? Is 2-dimethylpentane a correct name?

02 Write the structural formulas for the three isomeric saturated hydro-carbons having the molecular formula C_5H_{12}. Name each by the IUPAC system. Which one of these isomers would show a single peak in its proton NMR spectrum?

03 What is an alkyl group? What are the general molecular formulas for the alkyl groups derived from an alkane and a cycloalkane? How is the name of an alkyl group derived from the name of an alkane?

04 One objective of a satisfactory system of nomenclature is to give each different molecular structure a specific name; a correct name should describe a specific molecular structure. In what respect are the names "butene" and "butadiene" deficient?

05 Draw the molecular structures of the following compounds:

(a) 3-hexyne; (b) 1,3-pentadiene; (c) cyclobutene;
(d) 3,4-diethylhexane; (e) 1-butyne; (f) 2-methylpropene;
(g) 2-ethyl-3-methyl-1-butene; (h) 3-methyl-1-butyne; (i) *p*-dinitrobenzene;
(j) *n*-propylbenzene; (k) 1,3,5-tri-bromobenzene; (l) 1,3-diphenylbutane.

06 What type of isomerism is not disclosed in the name 2-hexene? How is the name for 2-hexene modified to show this isomerism? What is a polari-meter? To what do the terms levorotatory and dextrorotatory refer?

07 The molecular structure shown has the indicated IUPAC name with question marks where there should be numbers.

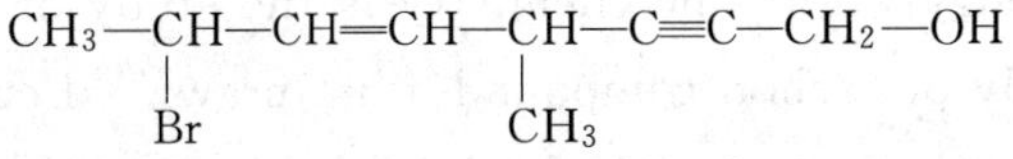

$$CH_3-\underset{Br}{CH}-CH=CH-\underset{CH_3}{CH}-C\equiv C-CH_2-OH$$

?-bromo-?-methyl-?-octen-?-yn-?-ol

Write the name by putting the correct numbers where question marks now appear.

08 The structures of many of the classes of compounds discussed in the class can be "derived" from that of the water molecule, H_a-O-H_b, by replacing one or both of the hydrogens by various organic groups. Name the class of compound formed when (a) H_a is replaced by an alkyl group, (b) H_b is replaced by an aromatic group, (c) H_a and H_b are replaced by alkyl groups, (d) H_a is replaced by the R—CO— group, (e) H_a is replaced by the R—CO— group and H_b is replaced by an alkyl group, and (f) H_a and H_b are replaced by R—CO— groups.

09 Write the structural formula for each of the following:

(a) 1,4-dichlorobenzene; (b) 3-hexyl-1-ol; (c) methoxycyclobutane;
(d) *p*-bromotoluene; (e) diethylamine; (f) *o*-methoxybenzaldehyde;
(g) propynal; (h) 3-hexanone; (i) 2-aminopropanoic acid;
(j) 2-methylbutanoic anhydride; (k) 3,5-dinitrobenzoyl chloride; (l) *N*,*N*-dimethylacetamide;
(m) *p*-phenylbenzamide; (n) polystyrene; (o) polyoxymethylene;
(p) poly (methylene); (q) nylon-6 and -6,6.

10 Using A and B for two different types of monomers, write structures which show how they might combine to form copolymers.

11 Describe the structure of an amino acid. What kind of isomerism do most amino acids exhibit? Why?

第 8 章 碳水化合物、氨基酸、蛋白质和脂肪
Chapter 8 Carbohydrates, Amino Acids, Protein and Fat

First, let us hear what Mike Adams, a science writer, is to say about organic chemistry and biochemistry: "Organic chemistry is the study of carbon compounds, and biochemistry is the study of carbon compounds that crawl." Let's further take a closer look at a typical fast-food dinner. Our food is a mixture of many different kinds of substances, but the energy we need to run our bodies comes from three of them: carbohydrates (the source of 40%~50% of our energy), protein (11%~14%), and fat (the rest). Table 8.1 shows typical energy and mass values for a burger, a serving of fries, and a milkshake. In order to undersatand what happens to these substances when we eat them, you need to know a little more about their composition.

Table 8.1 Fast-Food Dinner (These energy and mass values are derived from the USDA Nutrient Database for Standard Reference)

	Energy		Total mass	Protein mass	Carbohydrate mass	Fat mass
Large double hamburger with condiment	540.2 cal	2260 kJ	226.0 g	34.3 g	40.3 g	26.6 g
Fried potatoes	663.0 cal	2774 kJ	255.0 g	7.2 g	76.9 g	38.1 g
Chocolate milkshake	355.6 cal	1488 kJ	300.0 g	9.2 g	63.5 g	8.1 g

8.1 Carbohydrates

Carbohydrate is a general name for sugars, starches, and cellulose. It derives from an earlier belief that these substances were hydrates of carbon, because many of them have the general formula $(CH_2O)_n$. Today chemists also refer to carbohydrates as **saccharides**, after the smaller units from which they are built. Sugars are monosaccharides and disaccharides. Starches and cellulose are polysaccharides. Carbohydrates serve many different functions in nature. For example, sugar and starch are important for energy storage and production in both plants and animals, and cellulose provides the support structure of woody plants.

The most important **monosaccharides** are sugars glucose, fructose, and galactose, isomers with the general formula $C_6H_{12}O_6$. Each of these sugars can exist in either of two ring forms or in an open-chain form (Figures 8.1 and 8.2). In solution, they are constantly shifting from one form to another. Note that glucose and galactose have aldehyde functional groups in the open-chain form and that fructose has a ketone functional group. Glucose and galactose differ only in the relative position of the —H and —OH groups on the fourth carbon from the top.

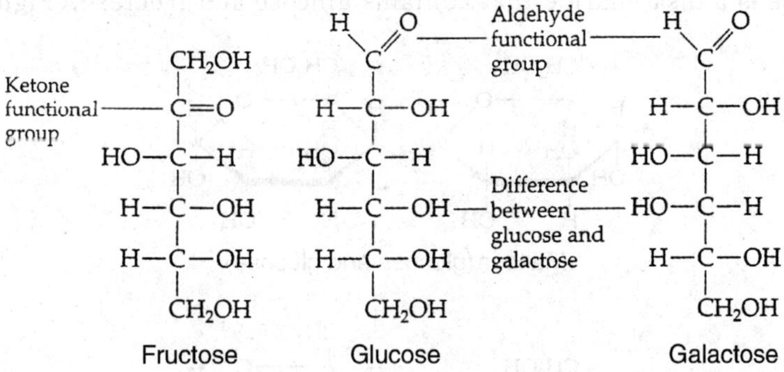

Figure 8.1 Open-Chain Form of Three Monosaccharides

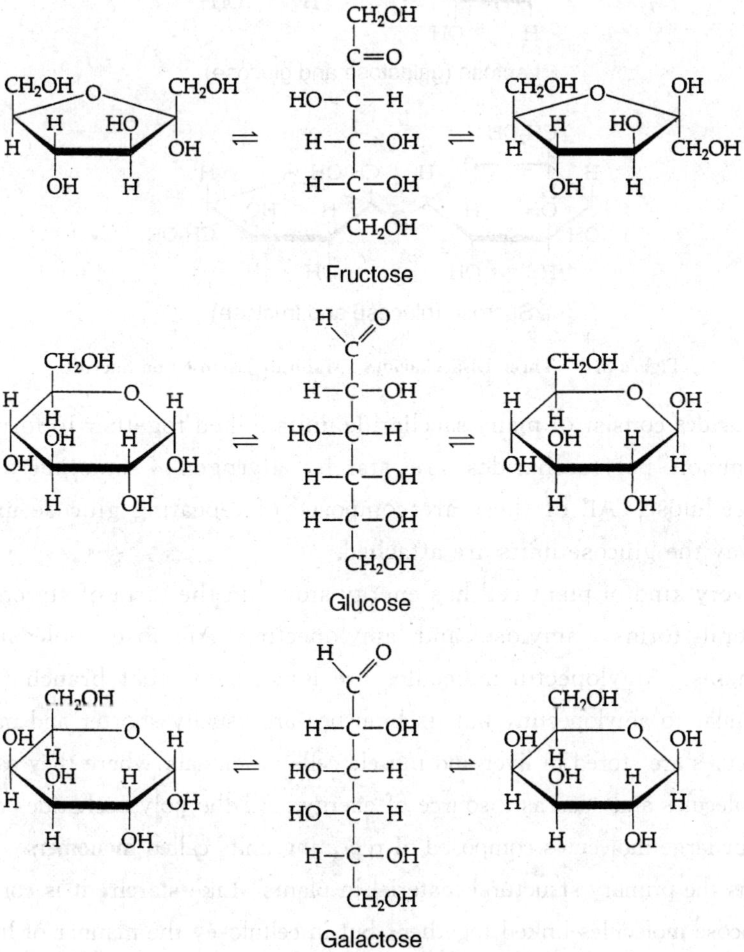

Figure 8.2 Fructose, Glucose and Galactose

Disaccharides are composed of two monosaccharide units. Maltose, a disaccharide consisting of two glucose units, is formed in the brewing of beer from barley in a process called malting. Lactose, or milk sugar, is a disaccharide consisting of galactose and glucose; sucrose is a disaccharide that contains glucose and fructose (Figure 8.3).

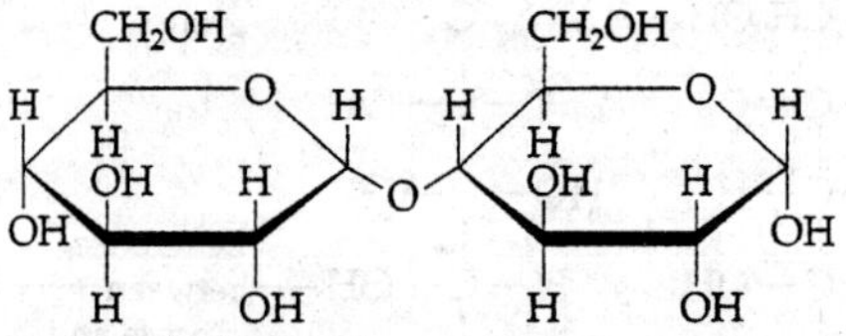

Maltose (glucose and glucose)

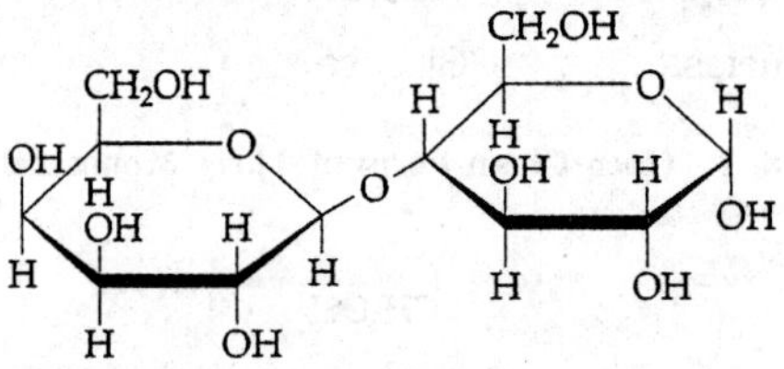

Lactose (galactose and glucose)

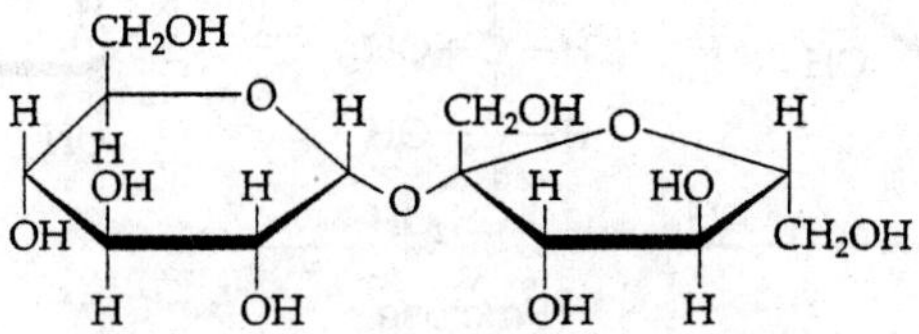

Sucrose (glucose and fructose)

Figure 8.3 Three Disaccharides: Maltose, Lactose and Sucrose

Polysaccharides consist of many saccharide units linked together to form long chains. The most common polysaccharides are starch, glycogen (sometimes called animal starch), and cellulose. All of these are composed of repeating glucose units, but they differ in the way the glucose units are attached.

Almost every kind of plant cell has energy stored in the form of starch. Starch itself has two general forms, amylose and amylopectin. Amylose molecules are long, unbranched chains. Amylopectin molecules are long chains that branch (Figure 8.4). Glycogen is similar to amylopectin, but its branches are usually shorter and more numerous. Glycogen molecules are stored in liver and muscle cells of animals, where they can be converted into glucose molecules and used as a source of energy. All the polysaccharides are **polymers**, a general name for large molecules composed of repeating units called **monomers**.

Cellulose is the primary structural material in plants. Like starch, it is composed of large numbers of glucose molecules linked together; but in cellulose, the manner of linking produces very organized chains that can pack together closely, allowing strong attractions to form (Figure 8.4). The strong structures that result provide support and protection for plants. Our digestive enzymes① are able to break the linkages in starch to release energy-producing glucose, but they are unable to liberate glucose molecules from cellulose because they cannot break the linkages there. Cellulose passes through our digestive tract unchanged.

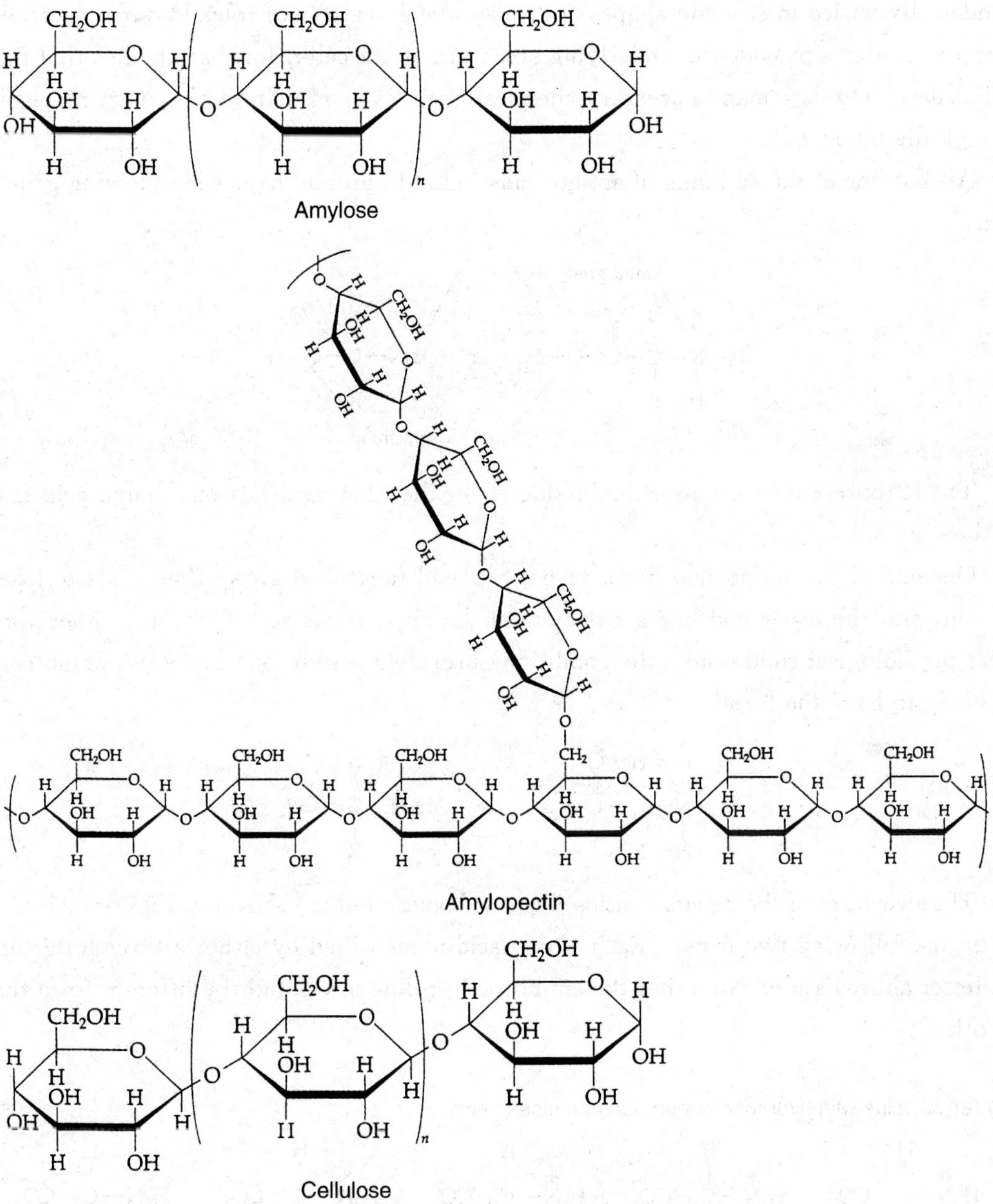

Figure 8.4 Polysaccharides

Starch provides energy both for plants and for the animals that eat the plants. Cellulose is an indigestible polysaccharide that provides structure for plants and fiber in animal diets.

---

① Remember that enzymes are naturally occurring catalysts—substances in plant and animal systems that speed chemical changes without being permanently altered themselves.

8.2 Amino Acids and Protein

Protein molecules are polymers composed of monomers called amino acids. Wonderfully varied in size and shape, they play a wide variety of roles in our bodies. For example, proteins provide the underlying structure of our cells, form antibodies that fight off invaders, regulate many necessary chemical changes, and help to transport molecules through the bloodstream.

All but one of the 20 kinds of amino acids found in protein have the following general form:

$$\text{Amine group} \diagdown \quad \diagup$$
$$H-\overset{H}{\underset{H}{\overset{|}{N}}}-\overset{}{\underset{R}{\overset{|}{C}}}-\overset{\ddot{O}}{\overset{\|}{C}}-\ddot{O}-H \quad \text{or} \quad H_2N-\overset{H}{\underset{R}{\overset{|}{C}}}-CO_2H$$
$$\diagup \quad \diagdown \text{Carboxylic acid group}$$

The R represents a group called a side chain that distinguishes one amino acid from another.

One end of the amino acid has a carboxylic acid functional group that tends to lose a H^+ ion, and the other end has a basic amine group that attracts H^+ ions. Therefore, under physiological conditions (the conditions prevalent within our bodies), amino acids are likely to have the form

$$H-\overset{H}{\underset{H}{\overset{|}{\overset{+}{N}}}}-\overset{}{\underset{R}{\overset{|}{C}}}-\overset{\ddot{O}}{\overset{\|}{C}}-\ddot{O}:^{-} \quad \text{or} \quad H_3\overset{+}{N}-\overset{H}{\underset{R}{\overset{|}{C}}}-CO_2^{-}$$

The structures of the 20 amino acids that our bodies need are shown in Figure 8.5(a)～(f) on the following two pages. Each amino acid is identified by either a three letter or a one-letter abbreviation. Note that the amino acid proline has a slightly different form than the others.

(a) Amino acids with hydrogen or hydrocarbon side chains

$$H_3\overset{+}{N}-\overset{H}{\underset{H}{\overset{|}{C}}}-CO_2^{-} \qquad H_3\overset{+}{N}-\overset{H}{\underset{CH_3}{\overset{|}{C}}}-CO_2^{-} \qquad H_3\overset{+}{N}-\overset{H}{\underset{\underset{CH_3}{|}}{\overset{|}{\underset{CH-CH_3}{C}}}}-CO_2^{-} \qquad H_3\overset{+}{N}-\overset{H}{\underset{\underset{\underset{CH_3}{|}}{\underset{CH-CH_3}{|}}}{\overset{|}{\underset{CH_2}{C}}}}-CO_2^{-} \qquad H_3\overset{+}{N}-\overset{H}{\underset{\underset{\underset{CH_3}{|}}{\underset{CH_2}{|}}}{\overset{|}{\underset{CH-CH_3}{C}}}}-CO_2^{-}$$

Glycine, Gly (G)　　Alanine, Ala (A)　　Valine, Val (V)　　Leucine, Leu (L)　　Isoleucine, Ile (I)

(b) Cyclic amino acid

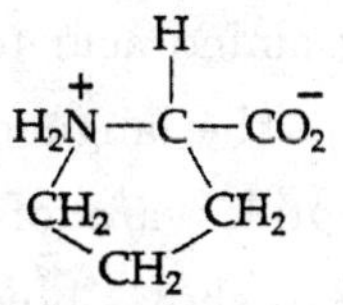

Proline, Pro (P)

(c) Aromatic amino acids

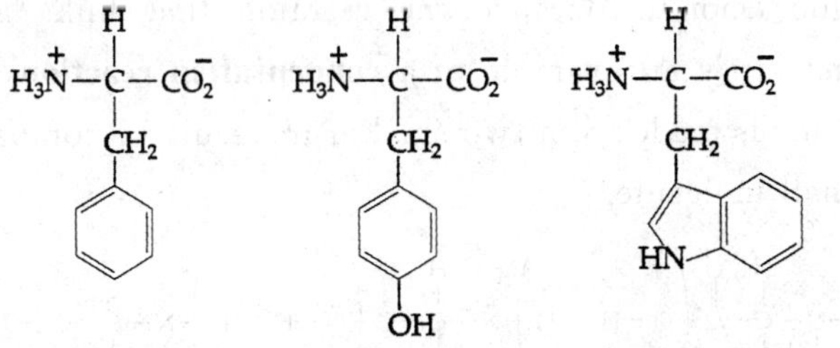

Phenylalanine, Phe (F)   Tyrosine, Tyr (Y)   Tryptophan, Trp (W)

(d) Amino acids with hydroxyl or sulfur containing side chains

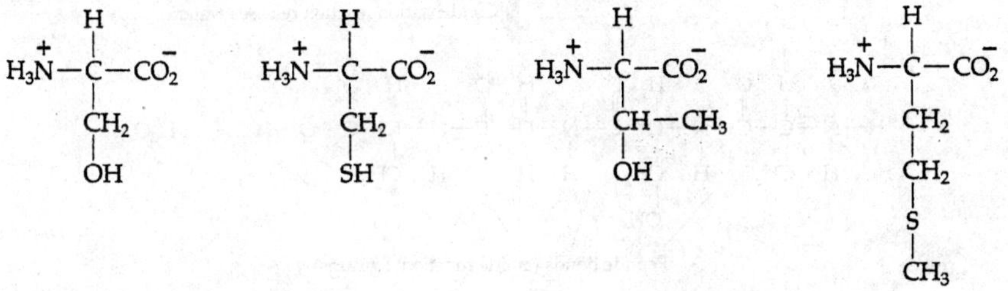

Serine, Ser (S)   Cysteine, Cys (C)   Threonine, Thr (T)   Methionine, Met (M)

(e) Basic amino acids

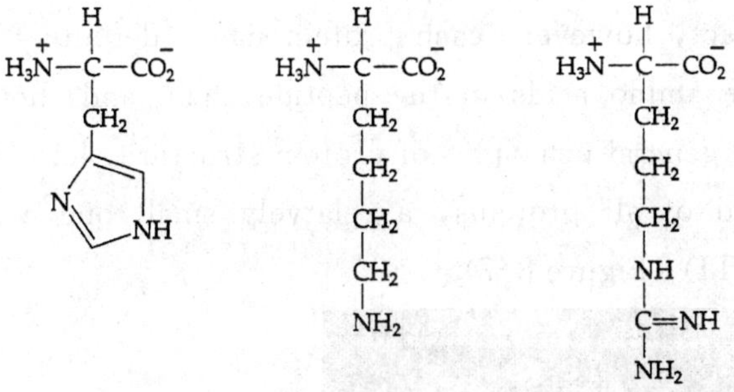

Histidine, His (H)   Lysine, Lys (K)   Arginine, Arg (R)

(f) Acidic amino acids and their amides

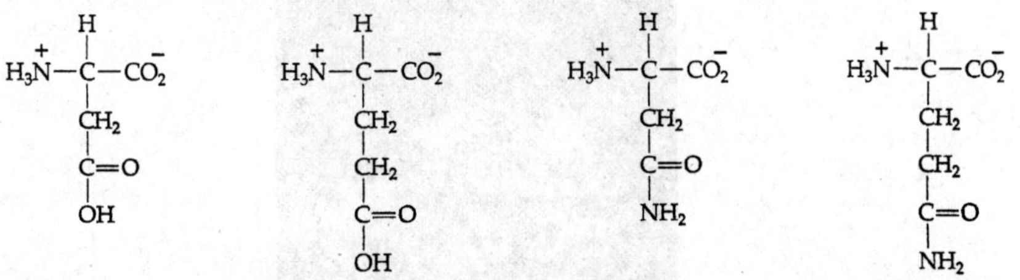

Aspartic acid, Asp (D)   Glutamic acid, Glu (E)   Asparagine, Asn (N)   Glutamine, Gln (Q)

Figure 8.5 Amino Acid Structures

Amino acids are linked together by a **peptide bond**, created when the carboxylic acid group of one amino acid reacts with the amine group of another amino acid to form an amide functional group. The product is called a **peptide**. Although the language used to describe peptides is not consistent among scientists, small peptides are often called **oligopeptides**, and large peptides are called **polypeptides**. Figure 8.6 shows how alanine, serine, glycine, and cysteine can be linked to form a structure called tetrapeptide (a peptide made from four amino acids). Because the reaction that links amino acids produces water as a by-product, it is an example of a **condensation reaction**, a chemical change in which a larger molecule is made from two smaller molecules accompanied by the release of water or another small molecule.

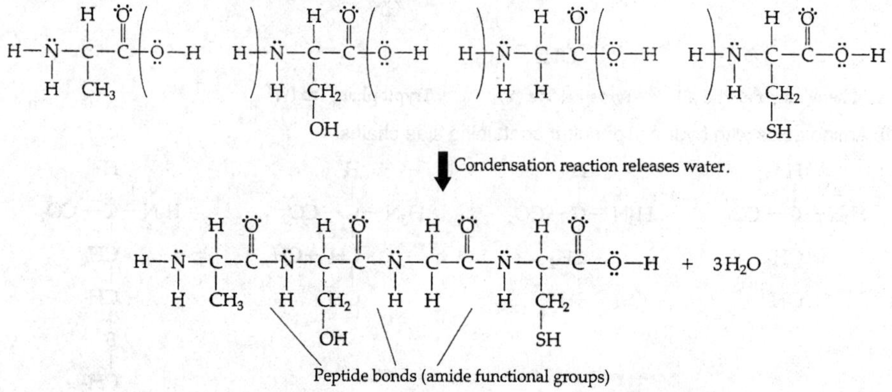

Figure 8.6 The Condensation Reaction that Forms the Tetrapeptide Ala-Ser-Gly-Cys

All **protein** molecules are polypeptides. At first glance, many of them look like shapeless blobs of atoms. In fact, however, each protein has a definite form that is determined by the order of the amino acids in the peptide chain and the interaction between them. To illustrate the general principles of protein structure, let's look at one of the most thoroughly studied of all proteins, a relatvely small one called bovine pancreatic trypsin inhibitor (BPTI) (Figure 8.7).

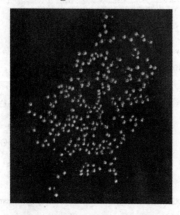

Figure 8.7 A Ball-and-Stick Model of a Protein Called Bovine Pancreatic Trypsin Inhibitor (BPTI)

Protein molecules are described in terms of their primary, secondary, and tertiary structures. The **primary structure** of a protein is the linear sequence of its amino acids. The primary strucure for BPTI is

Arg-Pro-Asp-Phe-Cys-Leu-Glu-Pro-Pro-Tyr-Thr-Gly-Pro-Cys-Lys-Ala-
Arg-Ile-Ile-Arg-Tyr-Phe-Tyr-Asn-Ala-Lys-Ala-Gly-Leu-Cys-Gln-Thr-
Phe-Val-Tyr-Gly-Gly-Cys-Arg-Ala-Lys-Arg-Asn-Asn-Phe-Lys-Ser-Ala-
Glu-Asp-Cys-Leu-Arg-Thr-Cys-Gly-Gly-Ala

The arrangement of atoms that are close to each other in the polypeptide chain is called the **secondary structure** of protein. Images of two such arrangements, an α-helix and a β-sheet are shown in Figures 8.8 and 8.9.

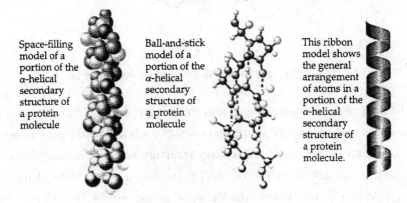

Figure 8.8 α-Helix

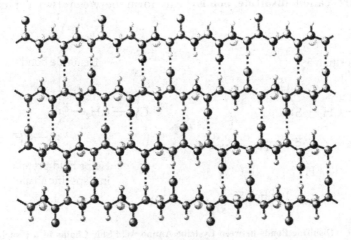

Figure 8.9 β-Sheet

BPTI contains both α-helix and β-sheet secondary structures, separated by less regular arrangements of amino acids. Because of the complexity of protein molecules, simplified conventions are used in drawing them to clarify their secondary and tertiary structures. Figure 8.10 shows the ribbon convention, in which α-helices are depicted by coiled ribbons and β-sheets are represented by flat ribbons.

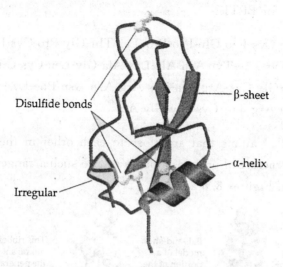

Figure 8.10　The Ribbon Structure of the Protein BPTI

When the long chains of amino acids link to form protein structures, not only do they arrange themselves into secondary structures, but the whole chain also arranges itself into a very specific overall shape called the **tertiary structure** of the protein. The protein chain is held in its tertiary structure by interactions between the side chains of its amino acids. For example, covalent bonds that form between sulfur atoms in different parts of the chain help create and hold the BPTI molecule's specific shape (Figure 8.10). These bonds, which are called **disulfide bonds**, can form between two cysteine amino acids (Figure 8.11).

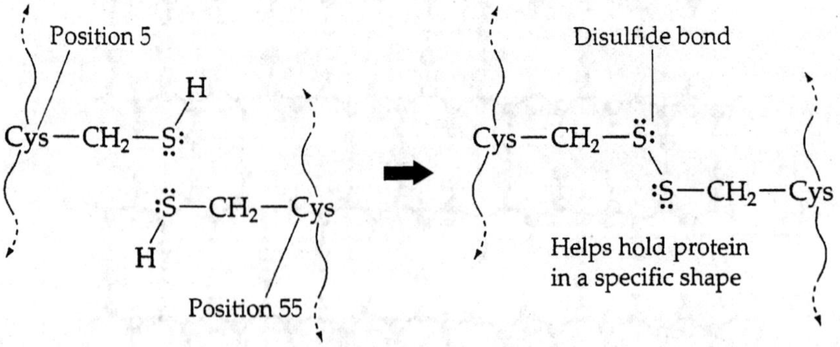

Figure 8.11　Disulfide Bonds between Cysteine Amino Acid Side Chains in a Protein Molecule

Hydrogen bonds can also help hold protein molecules in their specific tertiary shape. For example, the possibility of hydrogen bonding between their —OH groups will cause two serine amino acids in a protein chain to be attracted to each other (Figure 8.12).

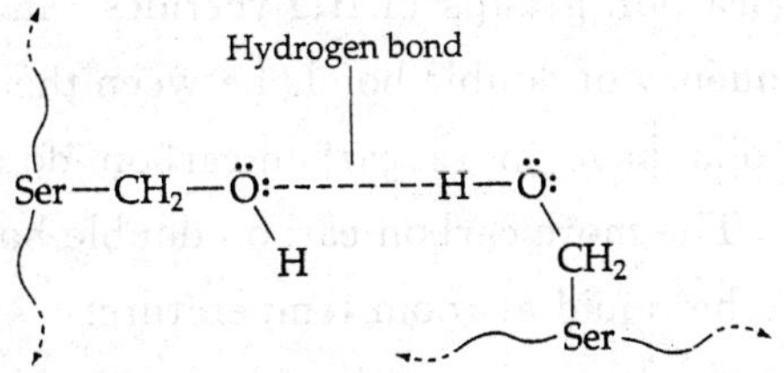

第 8 章 碳水化合物、氨基酸、蛋白质和脂肪(Carbohydrates, Amino Acids, Protein and Fat)

Figure 8.12 Hydrogen Bonding between Two Serine Amino Acids in a Protein Molecule

The tertiary structure is also determined by the creation of **salt bridges**, which consist of negatively charged side chains attracted to positively charged side chains (Figure 8.13).

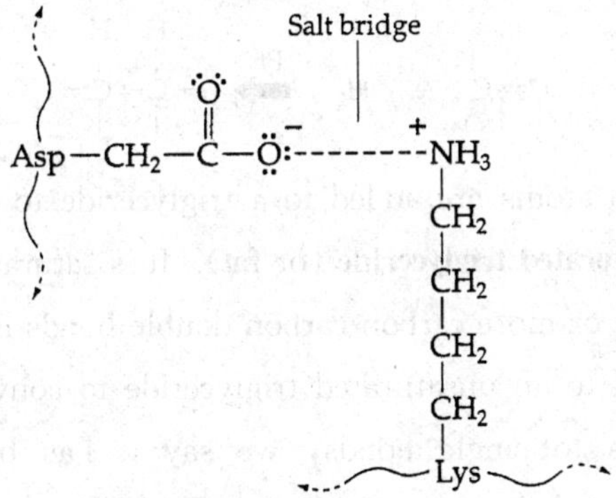

Figure 8.13 Salt Bridge between an Aspartic Acid Side Chain at One Position in a Protein Molecule and a Lysine Amino Acid Side Chain in Another Position

8.3 Fat

The fat stored in our bodies is our primary long-term energy source. A typical 70 kg human has fuel reserves of about 400 000 kJ in fat, 100 000 kJ in protein (mostly muscle protein), 2500 kJ in glycogen, and 170 kJ in total glucose. One of the reasons why it is more efficient to store energy as fat than as carbohydrate or protein is that fat produces 37 kJ/g, whereas carbohydrate and protein produce only 17 kJ/g.

As we saw before, animal fats and vegetable oils are made up of **triglycerides**, which have many different structures but the same general design: long-chain hydrocarbon groups attached to a three-carbon backbone.

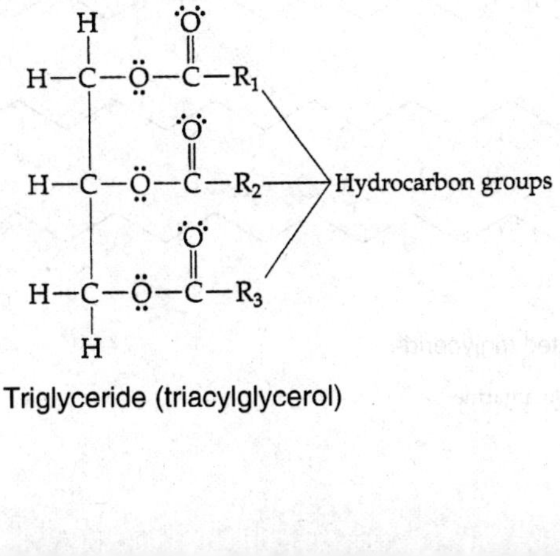

Triglyceride (triacylglycerol)

We saw that the hydrocarbon groups in triglycerides can differ in the length of the carbon chain and in the frequency of double bonds between their carbon atoms. The liquid triglycerides in vegetable oils have more carbon-carbon double bonds than the solid triglycerides in animal fats. The more carbon-carbon double bonds a triglyceride molecule has, the more likely it is to be liquid at room temperature.

A process called **hydrogenation** converts liquid triglycerides to solid triglycerides by adding hydrogen atoms to the double bonds and so converting them to single bonds. For example, the addition of hydrogen in the presence of a platinum catalyst changes corn oil into margarine.

$$\underset{/}{\overset{\backslash}{C}}=\underset{\backslash}{\overset{/}{C}} + H_2 \xrightarrow{Pt} -\underset{|}{\overset{H}{C}}-\underset{|}{\overset{H}{C}}-$$

When enough hydrogen atoms are added to a triglyceride to convert all double bonds to single bonds, we call it a **saturated triglyceride** (or **fat**). It is saturated with hydrogen atoms. A triglyceride that still has one or more carbon-carbon double bonds is an **unsaturated triglyceride**. If enough hydrogen is added to an unsaturated triglyceride to convert some but not all of the carbon-carbon double bonds to single bonds, we say it has been **partially hydrogenated**. Margarine is often described as being made with partially hydrogenated vegetable oils.

Unsaturated triglyceride
Liquid
Typical molecule in vegetable oil

$+ 5 H_2(g)$

Partially hydrogenated triglyceride
Solid
Typical molecule in margarine

Homework No. 8

01 Define the following terms.

    (a) carbohydrates;   (b) saccharides;   (c) monosaccharides;   (d) disaccarides;

    (e) polysaccharides;   (f) polymer;   (g) monomer;   (h) peptide bond;

    (i) peptide;   (j) condensation reaction;   (k) proteins;   (l) primary structure;

    (m) secondary structure;   (n) tertiary structure;   (o) triglyceride;   (p) hydrogenation;

    (q) saturated triglyceride;   (r) unsaturated triglyceride.

02 Complete the following statements by writing one of these words or phrases in each blank.

17 kJ/g; 37 kJ/g; acidic; active site; addition; amide; amino acids; amylopectin; amylose; benzen ring; carbon-carbon; cellulose; cholesterol; close to each other; condensation; diol; disaccharides; double bonds; energy; fatty acids; fructose; galactose; glucose; glucose units; glycerol; glycogen; hydrocarbon groups; hydrocarbons; hydrogenation; large; linear sequence; liver; long-term; monomers; monosaccharide; monosaccharides; muscle cells; n; —OH; overall shape; peptide; parenthese; partially; polysaccharides; protein; protein in our food; repeating units; shape; single; single bonds; small; small section; split in two; starches; step-growth; substrates; sugars; water.

    (a) Carbohydrate is a general name for _____, _____, and cellulose.

    (b) Sugars are monosaccharides and _____. Starches and cellulose are _____.

    (c) Disaccharides are composed of two _____ units.

    (d) Maltose is a disaccharide consisting of two _____ units.

    (e) Lactose, or milk sugar, is a disaccharide consisting of _____ and glucose.

    (f) Sucrose is a disaccharide that contains glucose and _____.

    (g) The most common polysaccharides are starch, _____ (sometimes called animal starch), and cellulose. All of these are composed of repeating _____, but they differ in the way the units are attached.

    (h) Almost every kind of plant cell has _____ stored in the form of starch. Starch itself has two general forms, _____ and _____.

    (i) Glycogen molecules are stored in _____ and _____ of animals, where they can be converted into glucose molecules and used as a source of energy.

    (j) All the polysaccharides are polymers, a general name for large molecules composed of _____ called monomers.

    (k) Our digestive enzymes are able to break the linkages in starch to release energy-producing glucose, but they are unable to liberate glucose molecules from _____ because they cannot break the linkages there.

    (l) Protein molecules are polymers composed of _____ called _____.

    (m) Amino acids are linked together by a(n) _____ bond, created when the carboxylic acid group of one amino acid reacts with the amine group of another amino acid to form a(n) _____ functional group.

    (n) A condensation reaction is a chemical change in which _____ or another small molecule is released.

    (o) The primary structure of a protein is the _____ of its amino acids.

(p) The arrangement of atoms that are _____ in the polypeptide chain is called the secondary structure of protein.

(q) The tertiary structure of a protein is its specific _____.

(r) The fat stored in our bodies is our primary _____ energy source.

(s) One of the reasons why it is more efficient to store energy as fat than as carbohydrate or protein is that fat produces _____, whereas carbohydrate and protein produce only _____.

(t) A process called _____ converts liquid triglyceridea to solid triglycerides by adding hydrogen atoms to the double bonds and so converting them to single bonds.

(u) When enough hydrogen atoms are added to a triglyceride to convert all double bonds to _____, we call it a saturated triglycedride (or fat). It is saturated with hydrogen atoms.

(v) A triglyceride that still has one or more carbon-carbon _____ is an unsaturated triglyceride.

(w) If enough hydrogen is added to an unsaturated triglyceride to convert some but not all of the carbon-carbon double bonds to single bonds, we say it has been _____ hydrogenated.

03 Identify each of the following structures as a carbohydrate, amino acid, peptide, triglyceride, or steroid.

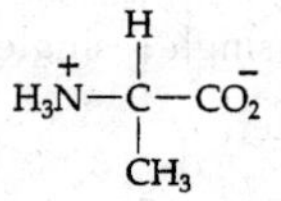

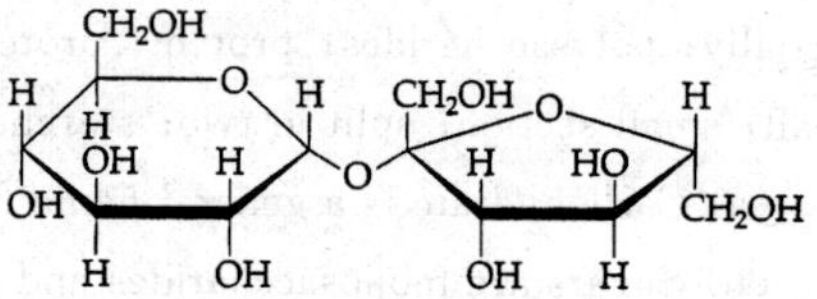

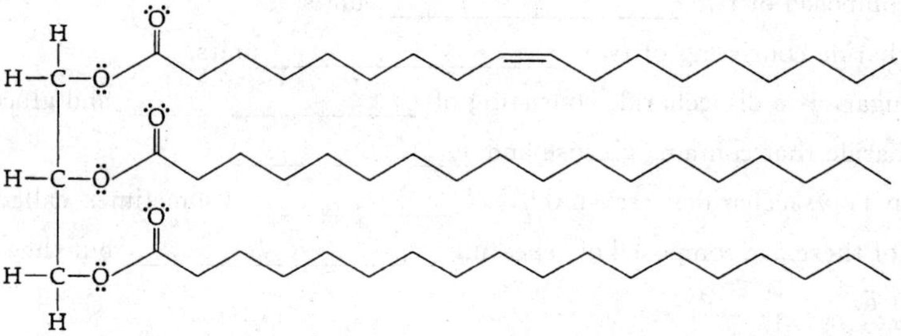

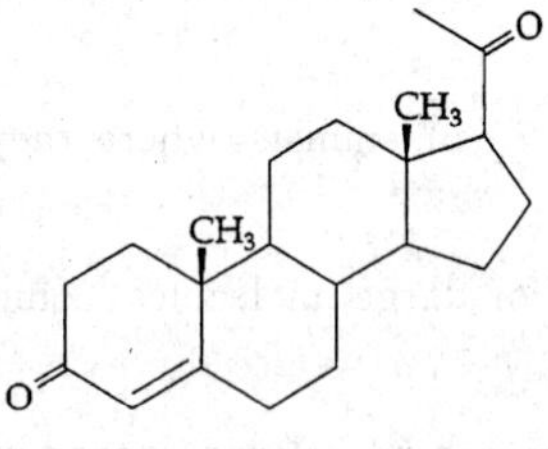

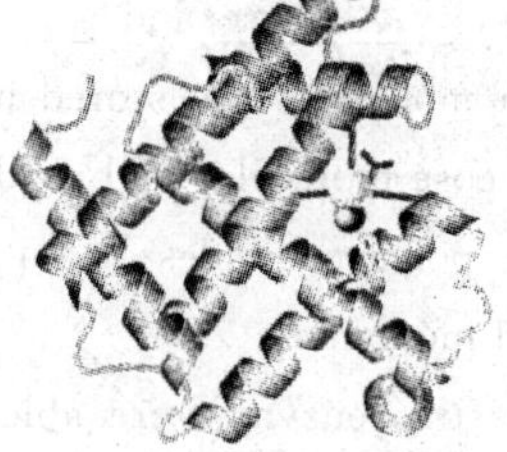

04 Identify each of the following structures as a monosaccharide, disaccharide, and polysaccharide.

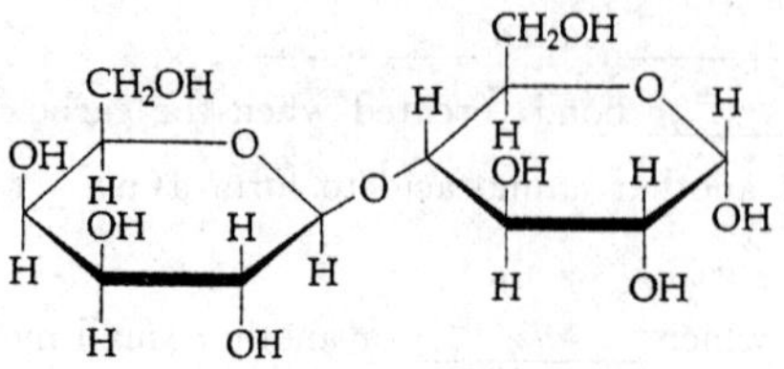

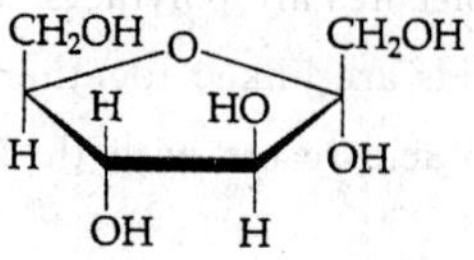

(c)
```
     CH₂OH
      |
      C=O
      |
  HO—C—H
      |
   H—C—OH
      |
   H—C—OH
      |
     CH₂OH
```

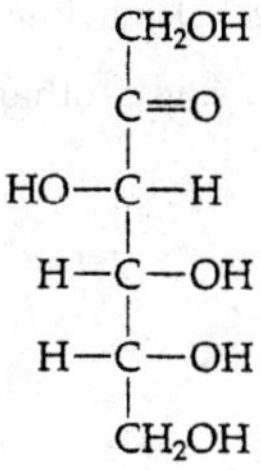

(d) [structure of amylose/cellulose polysaccharide chain]

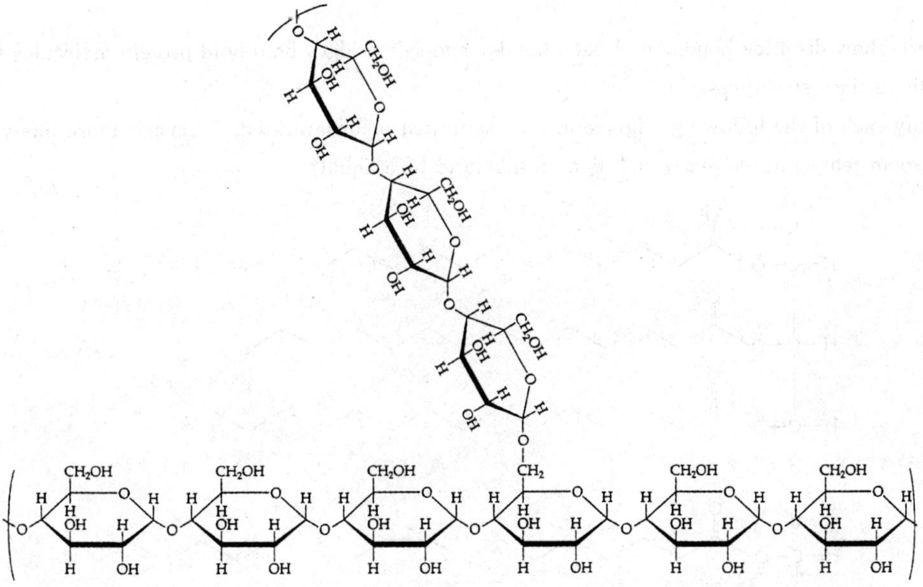

05 Identify each of the following as a monosaccharide, disaccharide, and polysaccharide.

  (a) maltose; (b) fructose; (c) amylose; (d) cellulose.

06 How do glucose and galactose differ?

07 What saccharide units form maltose, lactose, and sucrose?

08 Describe the similarities and differences between starches (such as amylose, amylopectin, and glycogen) and cellulose.

09 Explain why glycine amino acid molecules in our bodies are usually found in the second form shown below rather than in the first.

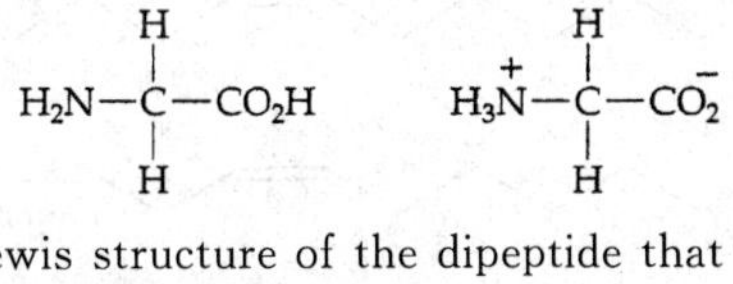

$$\begin{array}{cc} \text{H} & \text{H} \\ | & | \\ \text{H}_2\text{N—C—CO}_2\text{H} & \text{H}_3\overset{+}{\text{N}}\text{—C—CO}_2^- \\ | & | \\ \text{H} & \text{H} \end{array}$$

10 Using Figure 8.5, draw the Lewis structure of the dipeptide that has alanine combined with serine. Circle the peptide bond in your structure.

11 Show how the amino acids leucine, phenylalanine, and threonine can be linked together to form the tripeptide leu-phe-thr.

12 When the artificial sweetener aspartame is digested, it yields metahnol as well as the amino acids aspartic acid and phenylalanine. Although methanol is toxic, the extremely low levels introduced into the body by eating aspartame are not considered dangerous, but for people who suffer from phenylketoburia (PKU), the phenylalanine can cause severe mental retardation. Babies are tested for this disorder at birth, and when it is detected, they are placed on diets that are low in phenylalanine. Using Figure 8.5, identify the portions of aspartame's structures that yield aspartic acid, phenylalnine, and methanol.

$$H_2N-\underset{\underset{\underset{OH}{C=O}}{\overset{H}{\underset{|}{CH_2}}}}{\overset{H}{C}}-\overset{O}{\underset{\|}{C}}-\underset{H}{\overset{H}{N}}-\underset{\underset{}{\overset{H}{\underset{|}{CH_2-C_6H_5}}}}{\overset{H}{C}}-\overset{O}{\underset{\|}{C}}-O-CH_3$$

13 Describe how disulfide bonds, hydrogen bonds, and salt bridges help hold protein molecules together in specific tertiary structures.

14 Identify each of the following triglycerides as saturated or unsaturated. Which is more likely to be a solid at room temperature, and which is more likely to be a liquid?

15 Draw the structure of the triglyceride that would form from the complete hydrogenation of the following triglyceride.

第9章 聚合反应：由简单分子到高分子
Chapter 9　Polymerization Reaction: from Simple Molecules to Polymers

Polical events of the 1930s created an interesting crisis in fashion. Women wanted sheer stockings, but with the growing unrest in the world, manufacturers were having an increasingly difficult time obtaining the silk necessary to make them. Chemistry came to the rescue.

If you were a chemist trying to develop a substitute for silk, your first step would be to find out as much as you could about its chemical structure. Silk is a polyamide (polypeptide), a long-chain molecule (polymer) composed of amino acids linked together by amide functional groups (peptide bonds). Silk molecules are 44% glycine (the simplest of the amino acids, with a hydrogen for its distinguishing side chain) and 40% alanine (another very simple amino acid, with a —CH_3 side chain). Having acquired this information, you might decide to try synthesizing a simple polypeptide of your own.

The next step in your project would be to plan a process for making the new polymer, perhaps using the process of protein formation in living organisms as a guide. We saw before that polypeptides form in nature when the carboxylic acid group of one amino acid reacts with the amine group of another amino acid to form an amide functional group called a peptide bond. The reason why amino acids are able to form long chains in this way is that amino acids are difunctional: Each amino acid possesses both an amine functional group and a carboxylic acid functional group. After two amino acids are linked by a peptide bond, each of them still has either a carboxylic acid group or an amine free to link to yet another amino acid (Figure 8.6).

9.1 Step-Growth Polymerization and Condensation Polymers

W. H. Carothers, working for E. I. Du Pont de Nemours and Company, developed the first synthetic polyamide. He found a way to react adipic acid (a di-carboxylic acid) with hexamethylene diamine (which has two amine functinal groups) to form long-chain polyamide molecules call nylon 66. (The first "6" in the "66" indicates the number of carbon atoms in each portion of the polymer chain that are contributed by the diamine, and the second "6" shows the number of carbon atoms in each portion that are contributed by the di-carboxylic acid.) The reactants are linked together by condensation reactions in which an —OH group removed from a carboxylic functional group combines with a —H from an amine group to form water, and an amide linkage forms between the reacting molecules (Figure 9.1). When small molecules, such as water, are released in the formation of a polymer, the polymer is called a **condensation** (or sometimes a **step-growth**) **polymer**, and such polymerization reaction is termed **step-growth polymerization**.

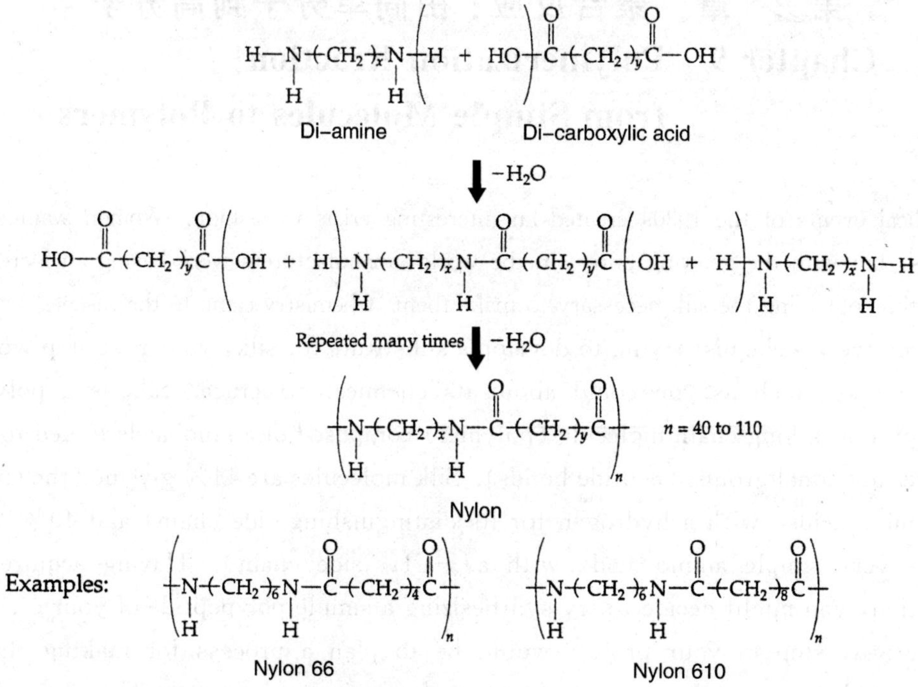

Figure 9.1 Nylon Formation

Chemists write chemical formulas for polymers by enclosing the repeating unit in parentheses followed by a subscript n to indicate that the unit is repeated many times:

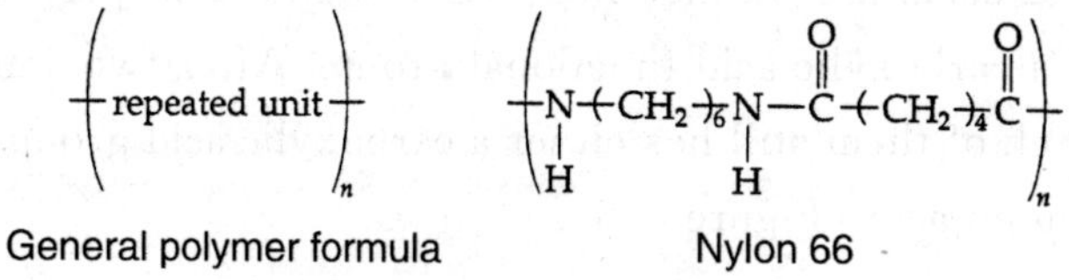

Nylon 66 was first made in 1935 and went into commercial production in 1940. Its fibers were strong, elastic, abrasion-resistant, lustrous, and easy to wash. With these qualities, nylon became more than just a good substitute for silk in stockings. Today nylon polymers are used in a multitude of products, including carpeting, upholstery fabrics, automobile tires, and turf for athletic fields.

One of the reasons for nylon's exceptional strength is the hydrogen bonding between amide functional groups in nylon's polymer structure, the stronger the hydrogen bonding between the chains. Thus changing the number of carbon atoms in the diamine (x in the Figure 9.1) and in the di-carboxylic acid (y in the Figure 9.1) changes the nylon's properties. For example, Nylon 610, which has four more carbon atoms in its dicarboxylic acid molecules than are found in Nylon 66, is somewhat weaker than Nylon 66 and has a lower melting point. Nylon 610 is used for bristles in paintbrushes.

Another example of condensation polymers is polyester similar to nylon, from which some movie stars' uniform and your own outfit are almost certainly made. Polyester are made from the reaction of a diol (a compound with two alcohol functional groups) with a di-carboxylic acid. Figure 9.2 shows the steps in the formation of poly(ethylenen terephthalate) from ethylene glycol and terephthalic acid.

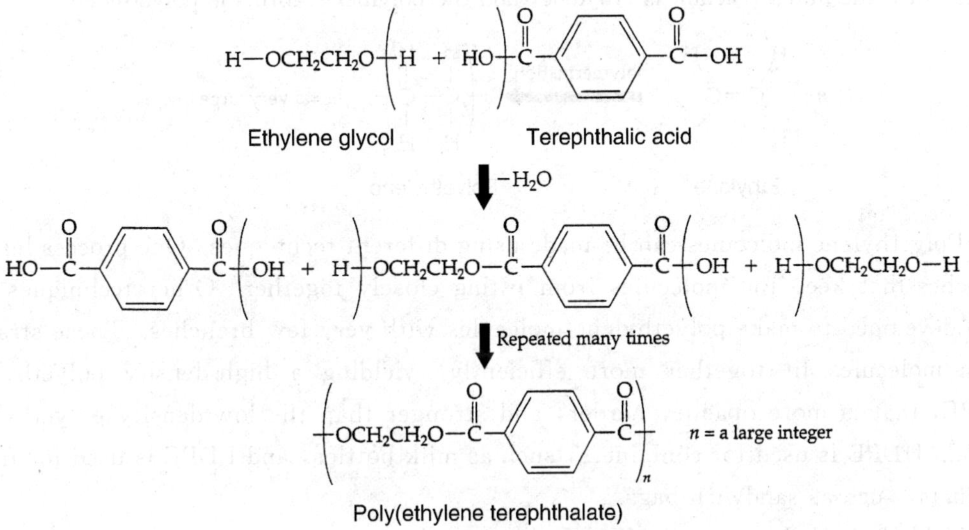

$$H-OCH_2CH_2O-H + HO-\overset{O}{\overset{\|}{C}}-\underset{}{\bigcirc}-\overset{O}{\overset{\|}{C}}-OH$$

Ethylene glycol   Terephthalic acid

↓ $-H_2O$

$$HO-\overset{O}{\overset{\|}{C}}-\underset{}{\bigcirc}-\overset{O}{\overset{\|}{C}}-OH + H-OCH_2CH_2O-\overset{O}{\overset{\|}{C}}-\underset{}{\bigcirc}-\overset{O}{\overset{\|}{C}}-OH + H-OCH_2CH_2O-H$$

↓ Repeated many times

$$\left(-OCH_2CH_2O-\overset{O}{\overset{\|}{C}}-\underset{}{\bigcirc}-\overset{O}{\overset{\|}{C}}-\right)_n \quad n = \text{a large integer}$$

Poly(ethylene terephthalate)

Figure 9.2 Polyester Formation

The transparency of which polyester is capable makes it a popular choice for photographic film and projection slides. Mylar, which is used to make long-lasting balloons, is a polyester, as is the polymer used for making eyeglass lenses. Polyesters have been used for fabrics that would once have been made from cotton, whose fundamental structure consists of the polymer cellulose. Polymer fibers, such as the fibers of Dacron© and Fortrel©—made from poly(ethylene terephthalate)—are about three times as strong as cellulose fibers, so polyester fabrics or blends that include polyester last longer than fabrics made from pure cotton. The strength and elasticity of polyesters make them ideal for sports uniforms.

9.2 Chain-Growth Polymerization and Addition Polymers

Unlike condensation (step-growth) polymers, which release small molecules, such as waters, as they form, the reactions that lead to **addition**, or **chain-growth**, **polymers** incorporate all of the reactants' atoms into the final product. Addition polymers are usually made from molecules that have the following general structure:

$$\underset{Y}{\overset{W}{\diagdown}}C=C\underset{Z}{\overset{X}{\diagup}}$$

Different W, X, Y, and Z groups distinguish one addition polymer from another.

(Visit the following web site to see one way in which addition polymers can be made. www.chemplace.com/college/)

If all of the atoms attached to the carbons of the monomer's double bond are hydrogen atoms, then the initial reactant is ethylene, and the polymer it forms is polyethylene.

$$n \;\; \underset{H}{\overset{H}{\diagdown}}C=C\underset{H}{\overset{H}{\diagup}} \xrightarrow{\text{polymerization}} \left(\begin{array}{c} H \;\; H \\ | \;\;\; | \\ -C-C- \\ | \;\;\; | \\ H \;\; H \end{array}\right)_n \qquad n = \text{a very large integer}$$

Ethylene   Polyethylene

Polyethylene molecules can be made using different techniques. One process leads to branches that keep the molecules from fitting closely together. Other techniques have been developed to make polyethylene molecules with very few branches. These straight-chain molecules fit together more efficiently, yielding a high-density polyethylene, HDPE, that is more opaque, harder, and stronger than the low-density polyethylene, LDPE. HDPE is used for containers, such as milk bottles, and LDPE is used for filmier products, such as sandwich bags.

Table 9.1 shows other addition polymers that can be made using monomers with different groups attached to the carbons in the monomer's double bond.

Table 9.1 Addition (Chain-Growth) Polymers

Initial Reactant	Polymer	Examples of Uses
Ethylene (H₂C=CH₂)	Polyethylene $-(CH_2-CH_2)_n-$	packaging, beverage containers, food containers, toys, detergent bottles, plastic buckets, mixing bowls, oil bottles, plastic bags, drapes, squeeze bottles, wire, and cable insulation
Propylene (H₂C=CHCH₃)	Polypropylene $-(CH_2-CH(CH_3))_n-$	clothing, home furnishings, indoor-outdoor carpeting, rope, automobile interior trim, battery cases, margarine and yogurt containers, grocery bags, caps for containers, carpet fiber, food wrap, plastic chairs, and luggage
Vinyl chloride (H₂C=CHCl)	Poly (vinyl chloride) $-(CH_2-CHCl)_n-$	"vinyl" seats in automobiles, "vinyl" siding for houses, rigid pipes, food wrap, vegetable oil bottles, blister packaging, rain coats, shower curtains, and flooring

续表

Initial Reactant	Polymer	Examples of Uses
Styrene	Polystyrene	foam insulation and packaging (Styrofoam®), plastic utensils, rigid, transparent salad containers, clothes hangers, foam cups, and plates

Homework No. 9

01 Define the following terms.

(a) condensation (or step-growth) polymer

(b) step-growth polymerization

(c) addition (or chain-growth) polymer

02 Complete the following statements by writing one of these words or phrases in each blank.

addition	amide	condensation	diol
monomers	n	—OH	parentheses
repeating unit	step-growth	water	

(a) The reactants that form nylon are linked together by _____ reactions in which an —OH group removed from a carboxylic functional group combines with a —H from an amine group to form water, and an amide linkage forms between the reacting molecules.

(b) When small molecules, such as water, are released in the formation of a polymer, the polymer is called a condensation (or sometimes _____) polymer.

(c) Chemists write chemical formulas for polymers by enclosing the repeating unit in _____ followed by a subscript _____ to indicate that the unit is repeated many times.

(d) Polyesters are made from the reaction of a(n) _____ (a compound with two alcohol functional groups) with a di-carboxylic acid.

(e) The reactions that lead to _____, or chain-growth, polymers incorporate all of the reactants' atoms into the final product.

03 Describe how Nylon 66 is made.

04 Describe the similarities and differences between the molecular structures of low-density polyethylene (LDPE) and high-density polyethylene (HDPE).

05 Explain why Nylon 66 is stronger than Nylon 610.

06 Identify each of the following as representing nylon, polyester, polyethylene, poly(vinyl chloride), polypropylene, or polystyrene. (In each case, the n represents some large integer.)

(a) $\left(\begin{array}{cc} H & H \\ | & | \\ -C-C- \\ | & | \\ H & H \end{array}\right)_n$

(b) $\left(-OCH_2CH_2O-\overset{O}{\overset{\|}{C}}-\underset{}{\bigcirc}-\overset{O}{\overset{\|}{C}}-\right)_n$

(c) $\left(\begin{array}{cc} H & H \\ | & | \\ -C-C- \\ | & | \\ H & CH_3 \end{array}\right)_n$

(d) $\left(\begin{array}{cc} H & H \\ | & | \\ -C-C- \\ | & | \\ H & C_6H_5 \end{array}\right)_n$

(e) $\left(-N(CH_2)_6N-\overset{O}{\overset{\|}{C}}(CH_2)_4\overset{O}{\overset{\|}{C}}-\right)_n$ with H on N's

(f) $\left(\begin{array}{cc} H & H \\ | & | \\ -C-C- \\ | & | \\ H & Cl \end{array}\right)_n$

07 Both ethylene and polyethylene are composed of nonpolar molecules. Explain why ethylene is a gas at room temperature while polyethylene is a solid at the same temperature.

第二部分（Part Ⅱ）

重要专业术语
(Significant Terms)

第10章 无机化学术语
Chapter 10　Inorganic Chemical Terms

1. periodic table, electronic structure

The periodic table groups the elements in order of increasing atomic number in such a way that elements with similar properties fall near each other. As the atomic number increases, the number of electrons in each atom also increases. A full appreciation of electronic structure—how the electrons are arranged in atoms—is essential for understanding the similarities and periodicities of the elements. Understanding electronic structure, in turn, requires a brief excursion into classical and modern physics. (The word "classical" is generally used for anything that was established and important in the past.)

2. wavelength, frequency, wave number, diffraction

Wavelength (λ, Greek lambda) is the distance between any two similar points on adjacent waves. The frequency (ν, Greek nu) of light is the number of complete waves, also known as the number of cycles, passing a given point in a unit of time. (Notice that the speed of light is equal to the product λ and ν, i.e., $c=\lambda\nu$.) The wave number is the number of wavelengths per unit of length covered, i.e., $\tilde{\nu}=1/\lambda$. Diffraction is the spreading of waves as they pass obstacles or openings comparable in size to their wavelength.

3. quantum, quantized, quantum theory, photoelectric effect, photon

Light can be regarded as made up of particles each of which carries a definite amount of energy, referred to as a quantum. Something that is quantized is restricted to amounts that are whole-number multiples of the basic unit, or quantum, for the particular system. Quantum theory is a general term for the idea that energy is quantized and the consequences of that idea. By assuming that light is quantized, Einstein was able to explain the photoelectric effect, in which electrons are released by certain metals (particularly Cs and the other alkali metals, Li, Na, K, and Rb) when light shines on them. (The photoelectric effect is used in practical devices such as automatic door openers.) A quantum of radiant energy is called a photon.

4. quantum mechanics, Heisenberg uncertainty priciple, momentum

"Mechanics" is the study of motion, and quantum mechanics refers to the study of the motion of entities that are small enough and move fast enough to have both observable wavelike and particlelike properties. What is called the Heisenberg uncertainty principle may be stated as follows: It is impossible to know simultaneously both the exact momentum and the exact position of an electron. (Momentum is mass times velocity. It expresses not only the tendency of a moving body to keep moving, but also, since velocity is a directional quantity, to maintain the direction of its motion.)

5. angular momentum, ground state, excited states, quantum number

Angular momentum, which is given by mass times velocity times the radius of a body's motion, is a measure of the tendency of a body to keep moving on a curved path. The lowest energy orbit, the one in which the single electron in a hydrogen atom normally resides, is the ground state for that electron. The states of energy higher than the ground state are excited states, reached by the electron when the atom has absorbed extra energy. A quantum number is a whole-number multiplier that specifies an amount of energy.

6. atomic orbital, the four quantum numbers

The region in which an electron with a specific energy will most probably be located is called an atomic orbital. The designation of the orbital "location" of an electron requires four quantum numbers. Independent of any experienced evidence, three of them arise from solutions of the Schrödinger equation. A fourth quantum number, the spin quantum number m_s ($=-1/2, +1/2$), is needed to complete the designation of each individual electron within an atom (because the electron can occupy the orbital in two different orientations). The first quantum number, the principal quantum number n, identifies the main energy levels (like the balconies). The second, the subshell quantum number l ($=0,1,2,\ldots,n-1$) (traditionally called either the angular momentum or azimuthal quantum number) identifies sublevels of energy within the main energy level (like the rows in each balcony). The third quantum number is the orbital quantum number m_l ($=-l,\ldots,0,\ldots,+l$) (traditionally called the magnetic quantum number)—it pins down the location of individual electrons in orbitals (like the seats in each row).

7. electron configuration, Pauli exclusion principle, Hund's principle

The electronic configuration of an atom is the distribution among the subshells of all of the electrons in the atom. Pauli exclusion principle: No two electrons can have the same quantum numbers. Hund's principle: Orbitals of equal energy are each occupied by a single electron before any of them acquires a second electron.

8. paramagnetism, diamagnetism

Paramagnetism is the property of attraction to a magnetic field shown by substances containing unpaired electrons. Diamagnetism is the property of repulsion by a magnetic field and shows the absence of unpaired electrons.

9. group (family), period, noble gas

The elements in a single vertical column in the periodic table are referred to as members of a group or family. A horizontal row in the periodic table is called a period. Each period ends with a noble gas—an element in which all energy sublevels that are occupied are completely filled.

10. representative elements, transition elements

The elements in which the s and p sublevels are filling are called the representative elements, which include alkali metals (Group I), alkaline earth metals (Group II), chalcogens (Group VI) and halogens (Group VII). The transition elements include all elements in which the d or f sublevels are filling. These are referred to as the d-transition elements and the f-transition elements. (The lanthanides and the actinides are f-transition elements.) Sometimes the f-transition elements are called the inner transition elements. Scandium, yttrium, and all of the sixth-period elements from lanthanum to lutetium are also known as the rare earth elements. Following uranium ($Z=92$) come the transuranium elements.

11. metals, nonmetals, semiconducting elements

Both in position in the periodic table and in properties, the semiconducting elements fall between the metals and the nonmetals.

12. chemical bond, valence electrons, Lewis symbol

In writing a definition of the chemical bond we must distinguish chemical bonding from other, weaker and less long-lasting forces. We choose to define the chemical bond as a force that acts strongly enough between two atoms or groups of atoms to hold them together in a different species that has measurable properties. Valence electrons are the electrons that are available to take part in chemical bonding. (The number of valence electrons in an atom of a representative element equals to the group number of that element.) In a Lewis symbol the outer electrons are indicated by dots (or circles, or x's, etc.) arranged around the atomic symbol. The pairing of two electrons in the same orbital is represented by two dots on the same side of the symbol.

13. chemical stability, octet rule, chemical reactivity

The noble gases fall at the end of each period in the periodic table and as a group are the least reactive of all the elements. This resistance to chemical change, or chemical stability, is credited to the completely filled outer s and p subshells of the noble gases. According to the octet rule, atoms tend to combine by gain, loss, or sharing of electron so that the outer energy level of each atom holds or shares four pairs of electrons. Sodium is a silvery metal that has high chemical reactivity, that is, a tendency to undergo chemical reactions.

14. metallic bonding, ionic bonding

Metallic bonding is the attraction between positive metal ions and surrounding, freely mobile electrons. Ionic bonding is the attraction between positive and negative ions.

15. Lewis structures, nonbonding electron pairs (lone pairs), covalent bonding, single, multiple (double, triple) and coordinate (donor atom and acceptor atom) covalent bond

Structures in which Lewis symbols are combined so that the bonding and nonbonding outer electrons are indicated are called Lewis structures. Pairs of valence electrons not involved in bonding are called nonbonding electron pairs, or lone pairs. Covalent bonding is based upon electron-pair sharing and is the attraction between two atoms that share electrons. A single covalent bond is a bond in which two atoms are held together by sharing two electrons. More than one pair of electrons can be shared between the same two atoms, resulting in what is called a mutiple covalent bond. In a double (triple) covalent bond two (three) electron pairs are shared between the same two atoms. A single covalent bond in which both electrons in the shared pair come from the same atom is called a coordinate covalent bond. The donor atom provides both electrons to a coordinate covalent bond, and the acceptor atom accepts an electron pair for sharing in a coordinate covalent bond.

16. resonance, resonance hybrid

Resonance refers to the arrangement of valence electrons in molecules or ions for which several Lewis structures can be written. The actual, single structure of a molecule or ion for which resonance structures can be written is called a resonance hybrid, since it has the characteristics of two or more of the possible structures.

17. nonpolar and polar covalent bond, dipole, network covalent substances

In molecules such as H_2, Cl_2, and N_2, the electron density (the probability of finding the valence electrons in a given area) is equally divided between the two bonded atoms. In a covalent bond of this type—a nonpolar covalent bond—the electrons are shared equally. A covalent bond in which electrons are shared unequally is called a polar covalent bond. A polar molecule is a dipole—a pair of opposite charges of equal magnitude at a specific distance from each other. The diamond and the other materials like it are network covalent substances—three-dimensional arrays of covalently bonded atoms.

18. bond length, bond dissociation energy, bond energy, lattice enegy

The distance between the two atoms at the point where their potential energy is at a minimum is the bond length—the distance between the nuclei of the two atoms in a stable molecule. Bond dissociation energy is the enthalpy per mole required to break exactly one bond of the same type per molecule. Bond energy is the average enthalpy per mole for breaking one bond of the same type per molecule.

The thermodynamic quantity that expresses the sum of the effects of the geometry of the arrangement of ions, the interionic distances, and the charges on the ions, and is used to measure the strength of ionic bonding is the lattice energy—the energy liberated when gaseous ions combine to give one mole of a crystalline ionic compound.

19. atomic radii, ionic radii, effective nuclear charge, screening effect, lanthanide contraction, isoelectronic ions

We use the term atomic radii to refer to an internally consistent set of radii for the elements based on the size of their atoms in single bonds. (These might also be referred to as covalent-metallic radii, for the values are derived from single-bonded covalent radii for nonmetals and metallic radii for metals.) Atomic radii are found by measuring the distances between bonded atoms and assigning part of the distance to each atom. Ionic radii are the radii of the anions and cations in crystalline ionic compounds. The effective nuclear charge is the portion of nuclear charge that acts on a given electron. The screening effect is the decrease in the nuclear charge acting on an electron due to the effects of other electrons. The gradual decrease in size across the first of the two f-transition elements series from lanthanum to lutetium is referred to as the lanthanide contraction. Isoelectronic ions of elements in the same period have identical electron configurations (e. g. Na^+ and Mg^{2+}).

20. ionization energy, noble gas configuration, electron affinity, d^{10} configuration, pseudo-noble gas configuration

The ionization energy (somtimes called the ionization potential) is the enthalpy change for the removal of the least tightly bound electron from an atom or an ion in the gaseous state (given per mole of atoms or ions of a given type). The removal of an electron from an atom or ion with a noble gas configuration is difficult, for this is a very stable configuration. The electron affinity is the enthalpy change for the addition of one electron to an atom or ion in the gaseous state (also given per mole of atoms or ions). A noble gas core with an outer d^{10} configuration, as in Ga^{3+}, is sometimes known as a pseudo-noble gas configuration.

21. polarization of an ion, electronegativity, electronegative atom, electropositive atom

The polarization of an ion is the distortion of its electron cloud by an ion of opposite charge. (In the presence of ions, the electron clouds of atoms and molecules can also be polarized.) Electronegativity is the ability of an atom in a covalent bond to attract electrons to itself. An electronegative atom tends to acquire a partial negative charge in a covalent bond or to form a negative ion. Nonmetals are generally electronegative. An electropositive atom tends to acquire a partial positive charge in a covalent bond or to form a positive ion. Metals are generally electropositive.

22. oxidation numbers, oxidation state

Oxidation numbers are assigned to atoms in compounds and are equal either to the charges of ions or to the charges atoms would have if the compound were ionic. The term oxidation state has the same meaning as "oxidation number."

23. molecular geometry, bond angle, bond axis, valence bond theory, σ bonds, hybridization, π bonds

Molecular geometry is the two- or three-dimensional arrangement in space of the atoms in a molecule. A bond angle is the angle between the bonds that joint two atoms to a third atom. In both valence bond theory and molecular orbital theory, nuclei are pictured as attracted to an area of high electron density located along the line between the two nuclei—the bond axis. At the same time, the bonding electrons are attracted by both nuclei. Valence bond theory describes bond formation as the interaction, or overlap, of atomic orbitals. All bonds in which the region of highest electron density surrounds the bond axis, are called σ bonds (sigma bonds). Hybridization is the mixing of the atomic orbitals on a single atom to give a new set of orbitals, called hybrid orbitals, on that atom. Only one σ bond can form between any two atoms. Multiple covalent bonds arise when p or d orbitals on atoms that are σ bonded to each other also overlap. π Bonds concentrate electron density above and below the bond axis and always have a plane of zero electron density passing through the bond axis. (π Bond formation can result from the interaction of p and p, p and d, and d and d orbitals.)

24. isomers, structural isomers, *cis-trans* isomerism (geometric isomerism), *cis* isomers, *trans* isomers

Compounds that differ in molecular structure but have the same molecular formula are called isomers. Structural isomers have the same molecular formula, but differ in how the atoms are bonded to each other. In *cis-trans* isomerism, or geometric isomerism, atoms or groups are arranged in different ways on either side of a double bond or some other rigid bond, such as that in a cyclic compound. In *cis* isomers the groups under consideration are on the same side of a double bond or other rigid structure. In *trans* isomers the groups under consideration are on opposite sides.

25. delocalized electrons, van der Waals forces, van der Waals radii, dipole moment, dipole-dipole interaction, London forces, hydrogen bond

Delocalized electrons are those electrons that occupy a space spread over three or more atoms. There are three principal types of intermolecular forces: dipole-dipole forces, hydrogen bonding, and London forces. Collectively, these are called van der Waals forces. The stronger the van der Waals forces, the higher the boiling points and heats of vaporization and fusion. The van der Waals radii are the radii of atoms not bonded to each other, but in contact at the most stable distance from each other. The degree of polarity of a molecule is measured by its dipole moment, μ (Greek mu, pronounced "muw"). The common unit for dipole moment is the debye, D (pronounced "de-buy", $1 D = 3.36 \times 10^{-30}$ C·m). In dipole-dipole interaction, molecules with dipole moments attract each other electrostatically; the positive end of one molecule attracts the negative end of another molecule, and so on, leading to an alignment of the molecules. London forces (also known as dispersion forces and being the result of momentary shifts in the symmetry of the electron cloud of a molecule) are the forces of attraction between fluctuating dipoles in atoms and molecules that are very close together. A hydrogen bond is the attraction of a hydrogen atom covalently bonded to an electronegative atom for a second electronegative atom.

26. nuclide, nucleons, nuclear force, nuclear chemistry, mass defect, nuclear binding energy, binding energy per nucleon, nuclear fusion, nuclear fission

Nuclide is a general term used to refer to any isotope of any element. Protons and neutrons, collectively called nucleons, are packed tightly together in the nucleus. The nuclear force is the force of attraction between nucleons; it acts between protons, between neutrons, and between protons and neutrons. Nuclear chemistry deals with nuclei and reactions that cause changes in nuclei. The difference between the sum of the masses of the individual nucleons and electrons and the mass of an atom is called the mass defect. The missing mass represents the nuclear binding energy—the energy that would be released in the combination of nucleons to form the nucleus. (The mass change due to the binding of the electrons is too small to be detected.) The binding energy per nucleon (also called average binding energy) is the nuclear binding energy of a nucleus divided by the number of nucleons in that nucleus. Nuclear fusion is the combination of two light nuclei to give a heavier nucleus. Nuclear fission is the splitting of a heavy nucleus into two lighter nuclei of intermediate mass numbers (and sometimes other particles as well).

27. radioactivity, radionuclides, magic number, half-life

Radioactivity is the spontaneous emission by unstable nuclei of particles, or electromagnetic radiation, or both. Nuclides that spontaneously break down, or decay, are called radioisotopes, radioactive nuclides, or radionuclides. Certain numbers of neutrons and protons—called magic numbers—impart particularly great nuclear stability. The magic numbers are 2, 8, 20, 28, 50, 82, and 126. Nuclides with magic numbers of both neutrons and protons are very stable and are abundant, for example, $_2^4$He, $_8^{16}$O, $_{20}^{40}$Ca, and $_{82}^{208}$Pb. The half-life of a radionuclide, $T_{1/2}$, is the time it takes for one-half of the nuclei in a sample of that nuclide to decay.

28. nuclear reactions, bombardment reactions, γ decay, α decay, β decay, antineutrino, neutrino, positron, electron capture

Nuclear reactions are the reactions that lead to changes in the atomic number, mass number, or energy states of nuclei. In bombardment reactions, electromagnetic radiation or fast-moving particles are captured by a nucleus to form an unstable nucleus that subsequently decays. α Decay and β decay are the major routes by which radioactive nuclides decay spontaneously. Frequently, α or β decay leaves the nucleus above its ground-state nuclear energy level. To return to this most stable energy level the nucleus undergoes γ decay—the emission of a γ ray—thereby giving up the excess energy in the form of electromagnetic radiation. The wavelength range for both X-rays and γ rays is from roughly 1 to 0.001 nm. X-rays are at the longer wavelength end of this range and γ rays at the shorter wavelength end of the range. However, the terms refer not to wavelength, but to the origin of the radiation—"X-rays" from energy changes of electrons and "γ rays" from nuclear energy changes. α Decay is the emission of an α-particle by a radionuclide. β Decay includes electron emission, positron emission, and electron capture. The antineutrino is a massless chargeless particle that is emitted with an electron, and the neutrino is a massless chargeless particle that is emitted with a positron. A positron is a particle identical to an electron in all of its properties except for the charge, which is $+1$ rather than -1. (Positrons were first detected during an investigation of cosmic rays.) The positive charge of an unstable nucleus can also be decreased by electron capture—the capture by the nucleus of one of its own inner orbital electrons.

29. chain reaction, critical mass, nuclear reactor, thermonuclear reactions, breeder reactor

A series of reactions in which the products of the reactions initiate other similar reactions, so that the series can be self-sustaining, is called a chain reaction. Nuclear fission becomes a self-sustaining chain reaction when the number of neutrons emitted equals or is greater than the number of neutrons absorbed by fissioning nuclei plus those lost to the surroundings. The critical mass of a fissionable material is the smallest mass that will support a self-sustaining chain reaction under a given set of conditions. The term nuclear reactor is usually applied to the equipment in which fission is carried out at a controlled rate. Nuclear reactions at very high temperatures (roughly $> 10^6$ K) are called thermonuclear reactions. A breeder reactor can produce at least as many fissionable atoms as it consumes.

30. hydration, solvation, chemical equilibrium, hydrolysis, hydrates, efflorescence, hygroscopic, deliquescent

The bonding of an ion or molecule to water molecules is called hydration. The bonding of the molecules or ions of dissolved substances to solvent molecules is solvation. In chemical equilibrium the rates of the forward and reverse chemical reactions are the same and the amounts of the species present do not change with time. Hydrolysis is the reaction in which the water molecule is split. Chemical compounds that include water molecules are called hydrates. The loss of water by a hydrate on exposure to air is called efflorescence. Some compounds are hygroscopic—they take up water from the air. Compounds that take up enough water from the air to dissolve in the water they have taken up are called deliquescent.

31. electrolytes, strong (weak) electrolytes, nonelectrolytes

Pure substances and substances which in solution conduct electricity by the movement of ions are called electrolytes. Compounds which are 100% dissociated or ionized in aqueous solution are called strong electrolytes. Acetic acid and other substances like it that are only partially ionized in aqueous solution are called weak electrolytes. Substances that dissolve in water to give solutions that do not conduct electricity are called nonelectrolytes.

32. acidic (alkaline) aqueous solution, acid, base, polyprotic acids, neutralization

An acidic (alkaline or basic) aqueous solution contains a greater concentration of H^+ (OH^-) ions than OH^- (H^+) of ions. An acid (water-ion) is a substance that contains hydrogen and yields hydrogen ions in aqueous solution. A base (water-ion) is a compound that contains hydroxide ions and when it dissolves in water dissociates to give hydroxide ions. Sulfuric acid and phosphoric acid, and other acids like them that contain more than one ionizable hydrogen atom, are known as polyprotic acids. Neutralization is the reaction of a strong acid with a strong base.

33. complex ion, ligands

A complex ion, or "complex", consists of a central metal atom or cation to which are bonded one or more molecules or anions. Transition metals, in particular, form complex readily. The molecules or ions bonded to the central atom or cation are called ligands.

34. hard water, carbonate hardness, water softening, non-carbonate hardness (permanent hardness), ion exchange

Hard water contains metal ions (principally Ca^{2+}, Mg^{2+} and Fe^{2+}) that form precipitates with soap or upon boiling. Water softening is the removal of the ions that cause hardness in water. Water that contains (both) Ca^{2+}, Mg^{2+} and Fe^{2+} ions but no (and) HCO_3^- ion has what is referred to as noncarbonate (carbonate) hardness or permanent (temporary) hardness. The process of replacement of one ion by another is known as ion exchange.

35. fossil fuels, oxidation, reduction, oxidation-reduction (redox) reactions, oxidizing agent, reducing agent

Hydrogen is the most abundant element in the cosmos and oxygen is the most abundant element in the crust of the earth. However, very little molecular hydrogen is present in the atmosphere of earth today. Hydrogen is known to be a component of the fossil fuels—coal, petroleum, and natural gas. Any process in which an oxidation number increases (decreases) algebraically is oxidation (reduction). Reactions in which oxidation and reduction take place are known as oxidation-reduction reactions, or frequently, as redox reactions. An atom, molecule, or ion that causes an increase (decrease) in the oxidation state of another substance and is itself reduced (oxidized) is called an oxidizing (reducing) agent.

36. heavy water, absorption

Deuterium, the isotope of mass number 2 of hydrogen, has a nucleus of one proton and one neutron and is represented by ^2_1H or D. The compound D_2O is the so called heavy water. Hydrogen is absorbed by many metals. Absorption is the incorporation of one substance into another at the molecular level.

37. acidic anhydride (oxide), basic anhydride (oxide), amphoteric

A nonmetal oxide such as $CO_2(g)$ and $SiO_2(s)$ that combines with water to give an acid is known as an acidic anhydride (anhydride meaning "without water"), or an acidic oxide. A metal oxide that yields a hydroxide base with water is known as basic anhydride, or a basic oxide, for example, $Na_2O(s)$ and $Fe_2O_3(s)$. Substances like aluminum oxide (Al_2O_3) that can act as either acids or bases are described as amphoteric.

38. allotropes

Ozone, O_3, is oxygen in the form of gaseous, triatomic molecules. Oxygen and ozone are allotropes—different forms of the same element in the same state (in this case, both gases). Most of the non-metals tend to "self-link" in this way—that is, to form chains. (The bonding of many atoms of the same element to each other in chains or rings is called catenation.)

39. Le Chatelier's principle

Le Chatelier's principle, which frequently applies to all chemical equilibria as well as to all equilibria of other types, is stated as follows: If a system at equilibrium is subjected to a stress, the system will react in a way that tends to relieve the stress.

40. acid salt, internal redox reaction, oxidizing anion, disproportionation reaction, oxidizing acids

An acid salt is formed by a metal cation and a hydrogen-containing anion from a polyprotic acid, e.g., $NaHSO_3$. An internal redox reaction is a reaction in which the oxidized and reduced elements originate in the same compound. An oxidizing anion is an anion capable of being reduced. In a disproportionation reaction an element in one oxidation state is both oxidized and reduced. Acids that are oxidizing agents are referred to as oxidizing acids.

41. Brønsted-Lowry acid, base, and acid-base reaction, oxo acids

Any molecule or ion that can act as a proton donor (acceptor) is a Brønsted-Lowry acid (base). A Brønsted-Lowry acid-base reaction is the transfer of a proton from a proton donor to a proton acceptor. Oxo acids, for example, H_2SO_4 or HNO_3, contain hydrogen, oxygen, and a third, central element.

42. equivalent mass of an acid or base, pH, pOH

The equivalent mass of an acid (base) is the mass of the acid (base) that donates one mole of H^+ (OH^-). Normality is the concentration of a solution expressed as the number of equivalents of solute per liter of solution. A solution containing 2 equivalents per liter is a 2 N (pronounced "two normal") solutions, and 0.5 L of a 2 N solution contains one equivalent of solute. The concentrations of hydrogen ion and hydroxide ion in aqueous solutions are usually expressed as pH and pOH, respectively, defined as $pH = -\lg[H^+]$ and $pOH = -\lg[OH^-]$. (N.B., The terms normality and equivalent mass are no longer in use.)

43. ion product constant for water, acid or base ionization constant

The ion product constant for water is $K_w = [H^+][OH^-]$. The acid and base ionization constants are $K_a = [H^+][A^-]/[HA]$ and $K_b = [BH^+][OH^-]/[B]$.

44. Lewis acid, base, and acid-base reaction

A Lewis acid is a molecule or ion that can accept one or more electron pairs. A Lewis base is a molecule or ion that can donate an electron pairs. A Lewis acid-base reaction is the donation of an electron pair from one atom to a covalent bond formed with another atom.

45. molecular orbital theory, molecular orbital, bonding or anti-bonding molecular orbital, bond order

Molecular orbital theory explains bond formation as the occupation by electrons of orbitals characteristic of bonded atoms rather than individual atoms. A molecular orbital is the space in which an electron with a specific energy is most likely to be found in the vicinity of two or more nuclei that are bonded together. A bonding molecular orbital is an orbital in which most of the electron density is located between the nuclei of the bonded atoms. An antibonding molecular orbital is an orbital in which most of the electron density is located away from the space between the nuclei. The bond order of a covalent bond between two atoms is the number of effective bonding electron pairs shared between the two atoms, i.e.,

$$\text{bond order} = \frac{(\text{the number of bonding electrons}) - (\text{the number of nonbonding electrons})}{2}$$

For example, the bond orders for H_2 and He are 1 and 0, respectively.

46. homonuclear, heteronuclear

Homonuclear means literally "the same nucleus"; the term refers to atoms of the same element. Heteronuclear means "of different nuclei"; that is, it refers to atoms of different elements.

47. complex, coordinate compound, coordinate number (complex), chelation, chelate ring, bidentate, tridentate, labile or inert complex

A complex is formed between a metal atom or ion that accepts one or more electron pairs, and ions or neutral molecules that contain nonmetal atoms which donate electron pairs. Any neutral compound that contains a metal atom and its associated ligands is called a coordination compound. The coordination number of the central metal atom or ion in a complex is the number of nonmetal atoms bonded to that atom or ion. The phenomenon of ring formation by a ligand in a complex is called chelation and the ring formed is called a chelate ring (pronounced "key-late", from the Greek *kela* meaning "crab's claw"). A ligand which contains two (three) donor atoms by which it can form a chelate ring (two fused rings) is referred to as bidentate (tridentate). A labile complex undergoes rapid exchange of its ligands with a reaction half-life of, say, a minute or less. By contrast, an inert complex has a slow rate of ligand exchange.

Symbols of Elements

Ac	Actinium	Gd	Gadolinium	Po	Polonium
Ag	Silver	Ge	Germanium	Pr	Praseodymium
Al	Aluminum	H	Hydrogen	Pt	Platinum
Am	Americium	He	Helium	Pu	Plutonium
Ar	Argon	Hf	Hafnium	Ra	Radium
As	Arsenic	Hg	Mercury	Rb	Rubidium
At	Astatine	Ho	Holmium	Re	Rhenium
Au	Gold	Hs	Hassium	Rf	Rutherfordium
B	Boron	I	Iodine	Rh	Rhodium
Ba	Barium	In	Indium	Rn	Radon
Be	Beryllium	Ir	Iridium	Ru	Ruthenium
Bh	Bohrium	K	Potassium	S	Sulfur
Bi	Bismuth	Kr	Krypton	Sb	Antimony
Bk	Berkelium	La	Lanthanum	Sc	Scandium
Br	Bromine	Li	Lithium	Se	Selenium
C	Carbon	Lr	Lawrencium	Sg	Seaborgium
Ca	Calcium	Lu	Lutetium	Si	Silicon
Cd	Cadmium	Md	Mendelevium	Sm	Samarium
Ce	Cerium	Mg	Magnesium	Sn	Tin
Cf	Californium	Mn	Manganese	Sr	Strontium
Cl	Chlorine	Mo	Molybdenum	Ta	Tantalum
Cm	Curium	N	Nitrogen	Tb	Terbium
Co	Cobalt	Na	Sodium	Tc	Technetium
Cr	Chromium	Nb	Niobium	Te	Tellurium
Cs	Cesium	Nd	Neodymium	Th	Thorium
Cu	Copper	Ne	Neon	Ti	Titanium
Db	Dubnium	Ni	Nickel	Tl	Thallium
Dy	Dysprosium	No	Nobelium	Tm	Thulium
Er	Erbium	Np	Neptunium	U	Uranium
Es	Einsteinium	O	Oxygen	V	Vanadium
Eu	Europium	Os	Osmium	W	Tungsten
F	Fluorine	P	Phosphorus	Xe	Xenon
Fe	Iron	Pa	Protactinium	Y	Yttrium
Fm	Fermium	Pb	Lead	Yb	Ytterbium
Fr	Francium	Pd	Palladium	Zn	Zinc
Ga	Gallium	Pm	Promethium	Zr	Zirconium

第11章 有机化学术语
Chapter 11 Organic Chemical Terms

11.1 Hydrocarbons

All organic compounds can be regarded as based on the structures of hydrocarbons. **Saturated hydrocarbons** contain only single covalent bonds; they may consist of straight or branched carbon chains (**alkanes**) or rings (**cycloalkanes**). All hydrocarbons with more than three carbon atoms exhibit structural isomerism. Groups containing one less hydrogen atom than an alkane are called **alkyl groups**. **Un-saturated hydrocarbons** are those containing double bonds (**alkenes**) or triple bonds (**alkynes**). Rotation of two carbon atoms joined by a double bond is inhibited by the π bond, and carbon atoms in rings are similarly unable to rotate. In both cases, *cis-trans* isomerism is therefore possible. A carbon-carbon double bond creates a planar region in a molecule. Compounds that contain **conjugated double bonds**—double bonds that alternate with single bonds—are often colored. **Aromatic hydrocarbons** are unsaturated compounds with planar ring systems stabilized by delocalized π bonding, as in benzene. Hydrocarbons that do not contain aromatic rings are termed **aliphatic**. The preferred system of nomenclature for organic compounds is that formulated by the International Union of Pure and Applied Chemistry, IUPAC.

Pairs of molecules with the same molecular formula that rotate **plane-polarized light** in opposite directions are called **optical isomers** or **enantiomers**. They are mirror images of one another and are usually very similar in their properties, and so are difficult to seperate. Optical isomers may arise when four different groups are bonded to one carbon atom, producing an **asymmetric** molecule. The property of having optical isomers is called **chirality**. A mixture of equal parts of the **levorotatory** and **dextrorotatory** isomers of a substance—a **racemic mixture**—causes no net rotation of polarized light.

The reactivity of saturated hydrocarbons is generally low, although hydrogen atoms are readily replaced by halogen atoms (**halogenation**). Double bonds or polar bonds provide sites where reaction is more likely to occur. Unsaturated hydrocarbons readily undergo **addition reactions**; aromatic compounds tend to undergo **substitution reactions**.

11.2 Hydrocarbons and Energy

The combustion of **fossil fuels**, which contain carbon and hydrocarbons, is our primary source of energy. **Natural gas** is a mixture of substances, with methane the chief component; it is recovered from oil wells and isolated natural gas wells. **Petroleum** is largely a mixture of hydrocarbons of various types. Ninety percent of all petroleum is burned for energy and ten percent is used in the manufacture of industrial organic chemicals, including plastics and rubber. Distillation is used to seperate crude petroleum into fractions of different boiling points and molecular masses. In the production of gasoline or other useful substances, these components are modifed in various ways (by **isomerization, cracking, alkylation,** and **reforming**). **Coal** contains carbon, hydrogen, nitrogen, oxygen, and sulfur in varying proportions. **Coke** is made by heating coal in the absence of air, a process that also yields ammonia and coal tar, from which other important organic compounds can be obtained. **Synfuels** are hydrocarbon fuels obtained by chemical synthesis or from natural sources by novel methods. They include liquid fuel made from coal, alcohol derived from biomass, hydrocarbons extracted from oil-bearing shale, substitute natural gas derived from coal, and methane produced by reaction of hydrogen with CO and CO_2. In the **Lurgi process**, coal reacts with steam and oxygen to produce a gas consisting of hydrogen, methane, carbon monoxide, and carbon dioxide. In **Fischer-Tropsch synthesis**, various hydrocarbons can be produced by the catalyzed reaction of hydrogen with carbon monoxide.

11.3 Some Introductory Concepts

A **functional group** is a chemically reactive atom or group of atoms that imparts characteristic properties to the family of organic compounds containing it. The site of reaction in an organic molecule is often a functional group, a multiple covalent bond, or a polar single bond. An electron-poor atom or group that will bond with an atom that has an available electron pair is called an **electrophile**. An electro-rich atom or group that will bond with an electron-deficient atom is called a **nucleophile.**

11.4 Organic Compounds and Their Formulas

Two co-workers at a pharmaceutical company, John and Stuart, jump into John's car at noon to drive four blocks to get some lunch. The gasoline that fuels the car is composed of many different organic compounds, including some that belong to the category of organic compounds called **alkanes** and a feul additive called methyl t-butyl **ether** (MTBE). When they get to the restaurant, Stuart orders a spinach and fruit salad. The spinach contains a **carboxylic acid** called oxalic acid, and the odor from the orange and pine-apple slices is due in part to the **aldehyde** 3-methylbutanal and the **ester** ethyl butanoate. The saland dressing is preserved with BHT, which is an **arene**. John orders fish, but he sends it back. The smell of the **amine** called trimethylamine let him know that it was spoiled.

The number of natural and synthetic organic, or carbon-based, compounds runs into the millions. Fortunately, the task of studying them is not so daunting as their number would suggest, because organic compounds can be categorized according to structural similarities that lead to similarities in compounds' important properties. For example, you know that alcohols are organic compounds possessing one or more —OH groups attached to a hydrocarbon group (a group that contains only carbon and hydrogen). Because of this structural similarity, all alcohols share certain chemical characteristics. Chemists are therefore able to describe the properties of alcohols in general, which is a great deal simpler than describing each substance individually.

In an introductory course on organic chemistry, you will learn how to recognize and describe alkanes, ethers, carboxylic acids, aldehydes, esters, arenes, amines, and other types of organic compounds.

Organic (carbon-based) compounds are often much more complex than inorganic compounds, so it is more difficult to deduce their structures from their chemical formulas. Moreover, many organic formulas represent two or more isomers, each with a Lewis structure of its own. The formula $C_6H_{14}O$, for example, has numerous isomers, including butyl ethyl ether, 1-hexanol and 3-hexonal.

Chemists have developed ways of writing organic formulas so as to describe their structures as well. For example, the formula for butyl ethyl ether can be written $CH_3CH_2CH_2CH_2OCH_2CH_3$, and the formula for 1-hexanol can be written $HOCH_2CH_2CH_2CH_2CH_2CH_3$ to show the order of the atoms in the structure.

Formulas like these that serves as a collapsed or condensed version of a Lewis structure are often called **condensed formulas** (even though they are longer than the molecular formulas). To simplyfy these formulas, the repeating —CH_2— groups can be represented by CH_2 in parentheses followed by a subscript indicating the number of times it is repeated. In this convention, butyl ethyl either becomes $CH_3(CH_2)_3OCH_2CH_3$, and 1-hexanol becomes $HOCH_2(CH_2)_4CH_3$.

The position of the —OH group in 3-hexanol can be shown with the condensed formulas $CH_3CH_2CH(OH)CH_2CH_2CH_3$. The parentheses, which are often left out, indicate the location at which the —OH group comes off the chain of carbon atoms. According to this convention, the group in parentheses is attached to the carbon that precedes it in the condensed formulas.

Although Lewis structures are useful for describing the bonding within molecules, they can be time-consuming to draw, and they do not show the spatial relationships of the atoms well. For example, the Lewis structure of butyl ethyl ether seems to indicate that the bond angles around each carbon atom are either 90° or 180° and that the carbon atoms lie in a straight line. In contrast, the ball-and-stick and space-filling models show that the angles are actually about 109° and that the carbons are in a zigzag arrangement. The highly simplified depiction known as a **line drawing** shows an organic structure's geometry better than a Lewis structure does and takes much less time to draw. Remember that in a line drawing, each corner represents a carbon, each line represents a bond (a double line is a double bond), and an end of a line without another symbol attached also represents a carbon. We assume that there are enough hydrogen atoms attached to each carbon to yield a total of four bonds.

Brief descriptions of some of the most important families of organic compounds lay a foundation for any study of organic chemistry. But this is not the place to describe the process of naming organic compounds, which is much more complex than naming inorganic compounds, except to say that many of the better-known organic substances have both a systematic and a common name. In the examples that follow, the first name presented reflects the rules set up by the International Union of Pure and Applied Chemistry (IUPAC). Any alternative names are given in parentheses. Thereafter, we will refer to the compound by whichever name is more frequently used by chemists.

11.5 Organic Chemistry as a Basis for Studying Biochemisty and Synthetic Polymers

It's Fridy night, and you don't feel like cooking, so you head for your favorite eatery, the local 1950s-style dinner. There you spend an hour talking and laughing with friends while downing a double hamburger, two orders of fries, and the thickest milkshake in town. After the food has disappeared, you're ready to dance the night away at a nearby club.

What's in the food that gives you the energy to talk, laugh, and dance? How do these substances get from your mouth to the rest of your body, and what happens to them once they get there? The branch of chemistry that answers these questions and many more is called biochemistry. The chemistry of biological systems. Because the scope of biochemistry is huge, we will attempt no more than a glimpse of it here by tracing some of the chemical and physical changes that food undergoes in your body. You will encounter the kinds of questions that biochemists ask and will see some of questions that they provide. Because chemicals that are important to biological systems are often organic or carbon-based compounds, we start our study of biochemistry with an introduction to organic chemistry.

It's not always apparent to the naked eye, but the structures of many plastics and synthetic fabrics are similar to the structures of biological substances. In fact, nylon was purposely developed to mimic the structural characteristics of protein, we will see how these substances are similar and how synthetic polymers are made and used.

第12章 物理化学术语
Chapter 12 Physical Chemical Terms

1. evaporation, vaporization, sublimation, condensation, liquefaction, melting point, fusion, (normal) boiling point, superheated, critical temperature, critical pressure, critical point, supercooled, triple point, normal freezing point

Evaporation is the escape of molecules from a liquid in an open container to the gas phase. Vaporization is the more general term for escape of molecules from the liquid or solid phase to the gas phase. Sublimation is the vaporization of a solid (the reverse transition, from the gas phase directly to the solid phase, is called "deposition"). Condensation is the movement of molecules from the gaseous phase to the liquid phase. We also speak of the transformation of a gas into a liquid as liquefaction. (Note the "e" in liquefaction—to use "i" in its place is wrong.) The melting point of a solid is the temperature at which the solid and liquid phases of a substance are at equilibrium. Fusion is a term also used in scientific publications to mean melting. (Remember that "fusion" means melting, not solidification.) The boiling point is the temperature at which the vapor pressure of a liquid equals the pressure of the gases above the liquid and bubbles of vapor form throughout the liquid. The normal boiling point is the boiling point at 760 mmHg, the atmospheric pressure at sea level. It is possible for a liquid to be superheated—heated to a temperature above the boiling point without the occurrence of boiling. Superheating occurs when it is difficult for molecules with enough kinetic energy to get together to form a bubble. The critical temperature is the temperature above which a substance cannot exist as a liquid no matter how great the pressure. The critical pressure is the pressure that will cause liquefaction of a gas at the critical temperature. The vapor-liquid equilibrium curve terminates at the critical point, at which the densities of the liquid and the vapor have become equal, and the boundary between the phases disappears. A liquid can be supercooled—cooled below its freezing point without the occurence of freezing. At the point where three lines intersect in a phase diagram—called a triple point—three phases are in equilibrium. The normal freezing point of a liquid is the temperature at which the liquid freezes at 760 mmHg pressure, that is, the temperature at which solid and liquid are in equilibrium at 760 mmHg pressure. The temperatures of the normal freezing point and the triple point are usually not the same.

2. viscosity, surface tension

The resistance of a substance to flow is viscosity—the opposite of fluidity. Surface tension is the property of a surface that imparts membrane-like behavior to the surface; it is formally defined as the amount of energy required to expand the surface of a liquid by a unit area.

3. crystalline soild, amorphous solid, crystal

A crystalline solid, also called a true solid, is a substance in which the atoms, molecules, or ions have a characteristic, regular, and repetitive three-dimensional arrangement. Sugar and salt are crystalline solids. An amorphous solid is a substance in which the atoms, molecules, or ions have a random and nonrepetitive three-dimensional arrangement. Tar is an amorphous solid. A crystal is a solid that has a shape bounded by plane surfaces intersecting at fixed angles. To a chemist, a crystal is an array of atoms, molecules, or ions in which a structural pattern is repeated periodically in three dimensions.

4. hexagonal or cubic closest packing, coordination number (crystal)

In the arrangement called hexagonal (cubic) closest packing, closest packed layers of atoms are arranged in an ABABAB....(ABCABCABC...) sequence. In both hexagonal and cubic packing, each sphere touches six other spheres in its own layer, plus three in the layer above and three in the layer below. This gives each sphere twelve nearest neighbors. The coordination number of an atom, ion, or molecule in a particular crystal structure is the number of nearest neighbors of that atom, ion, or molecule.

5. space lattice, crystal structure, unit cell, primitive unit cell, multiple unit cells, theoretical density

A space lattice is a system of points representing sites with identical environments in the same orientation in a crystal. The crystal structure of a substance is the complete geometrical arrangement of the particles that occupy the space lattice. A unit cell is the most convenient small part of a space lattice that, if repeated in three dimensions, will generate the entire lattice. A primitive unit cell is a unit cell in which only the corners are occupied. In some cases, unit cells are chosen that contain other lattice points in addition to those at the corners; these are called multiple unit cells. The mass of the unit cell of a metal divided by the volume of the unit cell gives the density of the metal, sometimes called the theoretical density.

6. polymorphous, radius ratio, isomorphous

Titanium dioxide (TiO_2) is an example of a compound that is polymorphous—able to crystallize in more than one crystal structure. The radius ratio—the ratio of cation to anion radii—needed for an ion to fit into a specific type of hole can be calculated from simple geometry. Substances that have the same crystal structure are said to be isomorphous. Some such substances can crystallize together to give a mixed product.

7. non-stoichiometric compounds

The tendency of some substances to incorporate point defects (variations in the occupation of the lattice or interstitial sites in the crystal) in their crystal structures accounts for the occurence of non-stoichiometric compounds—compounds in which atoms of different elements combine in other than whole-number ratios. Such compounds exist only in the solid state and in some cases have variable compositions. One example of such compounds is wustite, $Fe_{<1}O$.

8. saturated solution, solubility, unsaturated solution, super-saturated solution

In a saturated solution the concentration of dissolved solute is equal to that which would be in equilibrium with undissolved solute under the given conditions. The solubility of a substance is the concentration of a saturated solution (at 25°C if not otherwise stated). An unsaturated solution is one that can still dissolve more solute. A supersaturated solution holds more solute than would be in equilibrium with undissolved solute.

9. miscibility, infinitely miscible, immiscible

The mutual solubility of two substances in the same phase is known as miscibility. Substances that are similar in structure, bonding, and especially in intermolecular forces tend to be highly or even infinitely miscible ("like dissolves like"). A solid dissolves in a liquid if the interaction between the two substances is stronger than the forces holding the ions or molecules of the solid together. Liquids that are mutually insoluble, or very nearly so, are referred to as immiscible.

10. ideal solution of a molecular or an ionic solute, Henry's law

An ideal solution of a molecular solute is defined as one in which the forces between all particles of both solvent and solute are identical. An ideal solution of an ionic solute is defined as one in which the ions in solution are independent of each other and attracted only to solvent molecules. The relationship between the partial pressure of a solute gas and its solubility is expressed by Henry's law: At constant temperature, the partial pressure of a gas over a solution is directly proportional to the solubility of the gas in that solution.

11. standard solution, mass (weight) percent, molarity, molality

A solution of known concentration is called a standard solution. The concentration of a solution can be given in mass percent (weight percent). Mass % = [(mass of solute)/(mass of solution)]×100%. The expression of concentration in terms of the number of moles of solute per liter of solution is called molarity. Molarity = (moles of solute)/(liters of solution). Molality is defined as the number of moles of solute per kilogram of solvent). Molality = (moles of solute)/(kilogram of solvent).

12. Raoult's law, vapor pressure lowering, colligative property

Raoult's law can be stated as follows: The vapor pressure of a liquid in a solution is equal to the mole fraction of that liquid in the solution times the vapor pressure of pure liquid. The vapor pressure lowering is the difference between the vapor pressure of a pure solvent and the total vapor pressure over a solution of a nonvolatile solute. Any property of a solution that depends on the relative numbers of solute and solvent particles is called a colligative property (from the Latin *colligare*, to bind together).

13. semipermeable membranes, osmosis, osmotic pressure

Semipermeable membranes are membranes that allow the passage of some molecules but not others. The passage of solvent molecules through a semipermeable membrane from a more dilute solution into a more concentrated solution is called osmosis. Osmotic pressure is defined as the external pressure exactly sufficient to oppose osmosis and stop it.

14. colloidal dispersion, colloid, adsorption, sol, gel, emulsion, foam, aerosol

A mixture in which small particles remain dispersed almost indefinitely is called a colloidal dispersion, or simply a colloid. Strictly speaking, a colloid is a substance made up of suspended particles larger than most molecules, but too small to be seen in an optical microscopy. The attraction of a substance to the surface of a solid is called adsorption. An adsorbed layer can be held to a surface by intermolecular forces, the forces of chemical bonding, or a combination of the two. Mud or clay consists of solid particles suspended in a liquid—a type of colloid called a sol. A gel is a special type of colloid in which solid particles, usually very large molecules, unite in a random and intertwined structure that gives rigidity to the mixture. A colloid in which particles of a liquid are suspended in another liquid is called an emulsion. Foam is a colloid consisting of tiny bubbles of gas suspended in a liquid. Whipped cream is both an emulsion and a foam, for in it both butterfat and air bubbles are colloidally suspended. Smoke is a colloidal suspension of solid particles in air, and fog is a colloid consisting of suspended droplets of water in the air. A colloidal suspension in which air (or any gas) is the suspending medium is called an aerosol.

15. elementary reaction, reaction mechanism, intermediate

An elementary reaction is a reaction that occurs in a single step exactly as written. A reaction mechanism consists of all of the elementary steps in a single reaction pathway. An intermediate is a reactive species that is produced during the course of a reaction but always reacts further and is not among the final products of the reaction.

16. chemical kinetics, transition state (activated complex), free radical, activation energy

Chemical kinetics is the study of the rates and mechanism of chemical reactions. The short-lived combination of reacting atoms, molecules, or ions that is intermediate between reactants and products is called the transition state, or activated complex. A radical, or free radical, is a highly reactive species that contains an unpaired electron. The activation energy is defined as the minimum energy that reactants must have for reaction to occur. Customarily, activation energy is reported as the energy needed for the formation of one mole of the transition state.

17. molecularity, bimolecular or termolecular reaction

Elementary reactions are categorized in terms of their molecularity—the number of reactant particles involved in an elementary reaction. The reaction of nitric oxide and ozone is a bimolecular reaction—an elementary reaction that has two reactant particles.

The reactants in a bimolecular reaction can be the same or different—the point is that two particles must collide. A termolecular reaction is an elementary reaction that requires the interaction of three reactant particles.

18. reaction rate, initial reaction rate, rate equation, rate constant, overall reaction order, reactant reaction order, rate-determining step

Reaction rate is defined as the change in concentration of a reactant or a product in a unit of time. Rate = (change of concentration)/(time period of change). The instantaneous reaction rate at $t = 0$ s, the instant when the reaction begins, is called the initial reaction rate. A rate equation gives the mathematical relationship between the reaction rate and the concentration of one or more of the reactants. The proportionality constant k that relates reaction rate to some function of reactant concentrations raised to various powers is called the rate constant. The overall reaction order is defined as the sum of the exponents of the concentration terms in the rate equation. The reactant reaction order is the exponent on the term for that reactant in the rate equation. The rate-determining step is the slowest step in the reaction mechanism.

19. homogeneous or heterogeneous reaction or catalyst, physical or chemical adsorption, inhibitors, promoters

A homogeneous (heterogeneous) reaction is a reaction between substances in the same gaseous or liquid phase (different phases). Homogeneous catalyst are present in the same phase as the reactants. Heterogeneous catalyst are present in a phase different from that of the reactants and products. Adsorption is the adherence of atoms, molecules, or ions (the adsorbate) to a surface. It is called physical adsorption, or physisorption, when the forces between surface and adsorbate are van der Waals forces, and it is called chemical adsorption, or chemisorption, when the forces between adsorbate and surface are of the magnitude of chemical bond forces. Inhibitors are substances that slow down a catalyzed reaction—usually by tieing up the catalyst, possibly also by tying up a reactant. Promoters are substances that make a catalyst more effective.

20. equilibrium constant, reaction quotient

The equilibrium constant K is equal to the product of the concentrations of the reaction products, each raised to the power equal to its stoichiometric coefficient, divided by the product of the concentrations of the reactants, each raised to the power equal to its stoichiometric coefficient. The reaction quotient Q is a value found from an expression which takes the same form as the equilibrium constant but is used for reactions not at equilibrium.

21. hydrolysis of an ion, common ion (effect), buffer solution

The hydrolysis of an ion is the reaction of an ion with water to give either H_3O^+ or OH^-, plus whatever reaction product is formed by the ion. An ion added to a solution that already contains some of that ion is called a common ion. The common ion effect is a displacement of an ionic equilibrium by an excess of one or more of the ions involved. A buffer solution is a solution that resists changes in pH when small amounts of acid or base are added to it.

22. stoichiometric (equivalent) point, indicator, end point, titration curve

The stoichiometric point is the point at which the chemically equivalent amounts of reactants have reacted. An indicator is a compound that changes color in a specific pH range. The point at which an indicator changes color in a titration is usually referred to as the end point of the titration. A titration curve is a plot of pH versus volume of acid or base added.

23. solubility product, ion product, dissociation constants (complex ions)

The equilibrium constant for a solid electrolyte in equilibrium with its ions in solution is called the solubility product, or solubility product constant, K_{sp}. The ion product is equivalent to the reaction quotient for the dissolution of an ionic solid, Q_i. Equilibrium constants for the dissociation of complex ions are called dissociation constants, K_d.

24. entropy, absolute entropy, free energy change

Entropy S is a measure of disorder. The change in entropy of a system is a function both of the heat that flows into or out of the system and of the temperature. The absolute entropy is the entropy of a pure substance. The determination of absolute entropies is possible because there is a natural reference state for the entropy of a pure substance—the entropy at absolute zero. (It is impossible to determine absolute energy or enthalpy.) The free energy change, $G_2 - G_1$, is the energy that is available, or free, to do useful work as the result of a chemical or physical change.

25. half-reactions, ion-electron equations, redox couple

Every oxidation-reduction reaction can be divided into two half-reactions—reactions representing either oxidation only or reduction only. Ion-electron equations are balanced equations that include only the electrons and other species directly involved in oxidation or reduction of a given atom, molecule, or ion. The oxidized and reduced species that appear in an ion-electron equation are sometimes called a redox couple, or just a couple.

26. electrochemistry, electrochemical cell, cell reaction, half-cell, electrode reaction, voltaic cell, electrolytic cell, electrolysis, anode, cathode

Electrochemistry deals with oxidation-reduction reactions that either produce or utilize electrical energy. Any device in which an electrochemical reaction occurs is called electrochemical cell. The cell reaction is the overall chemical reaction that occurs in an electrochemical cell. Each electrode and its surrounding electrolyte make up a half-cell. The half-reaction that occurs in a half-cell is an electrode reaction. A voltaic cell generates electrical energy from a spontaneous redox reaction. An electrolytic cell uses electrical energy from outside the cell to cause a redox reaction to occur. The process of driving a non-spontaneous redox reaction to occur by means of electrical energy is called electrolysis. By definition, in any cell the anode is the electrode at which oxidation occurs, and the cathode is the electrode at which reduction occurs.

27. faraday, cell potential (electromotive force or emf), (standard) electrode potentials

One faraday is equivalent to 96 485 C (coulomb), or the amount of electrical charge represented by 1 mol of electrons. The cell potential (E or E_{cell}), also called the electromotive force or emf of a cell, is a measure of the potential difference between the two half-cells. An electrode potential is the potential for one half-reaction. The potential of 0.000 V for the H^+/H_2 couple and all other potentials established relative to it under the standard state conditions (all aqueous solutions at 1 mol/L concentrations, all gases at 1 atm partial pressure, and all pure solids and liquids in their most stable forms at 1 atm) are called standard electrode potentials, or standard reduction potentials.

28. overvoltage, storage batteries (accumulators, secondary cells), fuel cells, couple action

Overvoltage is a collective term for several effects that add to the voltage required by an electrochemical reaction. Batteries that can be recharged by using electrical energy from an external source to reverse the initial oxidation-reduction reactions are called storage batteries (or accumulators, or secondary cells). Fuel cells produce electrical energy directly from the air oxidation of a fuel continuously supplied to the cell. Corrosion is the blanket name applied to a large variety of deterioration processes undergone by manufactured materials when they experience loss and/or weakening during exposure to their environments. We are interested here in the corrosion of metals, which is an electrochemical phenomenon. If the connection between metal and the cathode is direct, the metal and the cathode together in their environment are called a couple and the corrosion is referred to as couple action.

第13章 分析化学术语
Chapter 13　Analytical Chemical Terms

The Importance of Analytical Chemistry

　　Historically, analytical chemistry has always occupied a vital position in the development of chemistry. The successful elucidation of the process of combustion by Lavoisier was due mainly to his employment of a balance in his investigations; he was among the first to recognize the immense power of quantitative measurements in chemical research. The atomic concept of matter dates back at least to ancient Greece, and certainly was not original with John Dalton. Dalton's contribution, above all, was to introduce a quantitative aspect to this notion—an aspect that was verifiable by actual experiment. In a very real sense, then, chemical analysis provided the support necessary to convert the atomic theory from a philosophical abstraction into something of physical significance.

　　Early chemistry was principally analytical in nature. Only as the body of experimental fact increased did it become possible for the chemist to specialize—according to his interests—in other fields. Irrespective of choice, however, he continued to rely heavily upon analytical methods and techniques to provide him with experimental information. Analytical chemistry thus assumed the supporting role of an indispensible tool in advancing the state of knowledge in the fields of inorganic, organic, and physical chemistry.

　　This situation is as applicable to the chemistry of today as to that of the past; every experimental investigation relies, to an extent, upon the results of analytical measurements. A thorough background in analytical chemistry is thus a vital necessity for all who aspire to be chemists, regardless of their field of specialization. Nor need these remarks be limited to prospective chemists. Investigators in virtually all of the physical and biological sciences are obliged to make use of analytical data in the course of their work. The physician relies heavily upon the results of analysis of body fluids in making his diagnoses. Analytical techniques are indispensible in the biochemist's study of living matter and its metabolic processes. The classification of a mineral is incomplete without knowledge of its chemical composition. Analytical techniques are employed by the physicist in identifying the products of high-energy bombardments. A catalogue such as this can be extended virtually without limit.

　　The employment of analytical chemistry in modern industry is of inestimable importance. It is difficult to imagine an article of present-day commerce whose raw materials have not, at some stage, been subjected to analytical control. The uniform quality of the paper upon which these words are printed is due in part to careful analysis during the various phases of its production; hundreds of analyses are performed upon the materials that go into as complex a commodity as an automobile.

Finally, aside from these highly practical considerations, a study of quantitative analysis is of benefit in that it places the highest premium upon careful, orderly work and intellectually honest observation; regardless of one's ultimate field of endeavor, these are traits worthy of cultivation.

1. analytical chemistry, qualitative or quantitative analysis

Analytical chemistry comprises the techniques and methods that provide answers to the questions "What?" and "How much?" with respect to the chemical composition of a sample of matter. The former is the province of qualitative analysis. Quantitative analysis is concerned with the problems attending the determination of the amount of species present in a given sample.

2. chemical analysis, instrumental analysis

Chemical analysis is based on chemical reactions while instrumental analysis relies upon optical, electrochemical, and other physical or physicochemical properties of sample solutions.

3. gravimetric analysis, volumetric analysis, colorimetric analysis, electroanalysis

The ultimate aim of a quantitative analysis is to ascertain how much of a given species is present in a sample of matter; depending upon the procedure employed, this may be accomplished directly or very indirectly. Regardless of how it is done, however, a final measurement of some sort is inherent in every determination and from this, the quantity of the species in question is derived. It is convenient to classify the methods of quantitative analysis according to the nature of this final measurement. Thus if this consists of securing the weight of a solid, the method is classed as a gravimetric analysis; where the final measurement involves determination of a volume, the method is called a volumetric analysis; if the absorption of light is measured, the procedure is sometimes termed a colorimetric analysis; and where an electrical property is determined, the method can be classified as electroanalytical.

4. mean (arithmetic mean, average), median

The mean, arithmetic mean, and average are synonymous terms that refer to the numerical value obtained by dividing the sum of a set of replicate measurements by the number of individual results in the set. The median of a set is that value about which all others are equally distributed, half being numerically greater and half being numerically smaller. If the set consists of an odd number of measurements, selection of the median may be made directly; for a set containing an even number of measurements, the average value of the central pair is taken as the median. We shall see that in the ideal case the mean and median are numerically equal; this fails to be true more often than not, however, when only a small set of measurements has been taken.

5. precision, absolute deviation, relative deviation, standard deviation

The term precision is frequently used to describe the reproducibility of results. It can be defined as the agreement between the numerical values of two or more measurements that have been made in an identical fashion. Absolute deviation is simply the difference between an experimental value and that which is taken as the best for the set (usually the arithmetic mean). Relative deviation is defined as average absolute deviation divided by the mean. The standard deviation is equal to the square root of the quantity obtained by the division of the sum of the squares of absolute deviations by the number of times of measurements minus one.

6. accuracy, absolute error, relative error

The term accuracy denotes the nearness of a measurement to its accepted value and is expressed in terms of error. Absolute error is the difference between the observed value and the accepted value, while relative error the division of the absolute error by the accepted value.

7. precipitation method, volatilization method

Two general types of gravimetric analyses are precipitation method and volatilization method. In the former, the substance to be determined is isolated from the other constituents in the sample by formation of an insoluble precipitate; the analysis is completed by determining the weight of this precipitate, or of some substance formed from it, by suitable treatment. The latter takes advantage of the property of volatility; here the substance to be determined is isolated by distillation. The product may either be collected and weighed, or the weight loss in the sample as a result of the distillation may be measured. Of the two, precipitation methods are the widely used.

8. interference

Compounds or elements that prevent the direct measurement of the species being determined are called interferences.

9. calibration

Probably the simplest method of calibration of analytical weights involves a direct comparison of each weight in a set with one whose value is known with certainty.

10. gravimetric method, volumetric method

A gravimetric method is one in which the analysis is completed by a weighing operation. A volumetric method is one in which the analysis is completed by measuring the volume of a solution of established concentration needed to react completely with the substance being determined. Ordinarily, volumetric methods are equivalent in accuracy to gravimetric procedures and are more rapid and convenient; their use is widespread.

11. titration, back-titration, standard solution, primary standard, standardization, end point, titration error, indicators

A titration is a process wherein the capacity of a substance to combine with a reagent is quantitatively measured. Ordinarily this is accomplished by the controlled addition of a reagent of known concentration to a solution of the substance until reaction between the two is judged to be complete; the volume of reagent is then measured. Occasionally it is convenient or necessary to carry out a volumetric analysis by adding an excess of the reagent and then determining the excess by titration with a second reagent of known concentration. The second titration is called a back-titration.

The reagent of exactly known composition used in a titration is called a standard solution. The accuracy with which its concentration is known sets a definite limit upon the accuracy of the method; for this reason, much care is taken in the preparation of standard solutions. Commonly the concentration of a standard solution is arrived in either of two ways: (1) A carefully measured quantity of a pure compound is titrated with the reagent and the concentration calculated from the weight and volume measurements; or (2) The standard solution is prepared by dissolving a carefully weighed quantity of the pure reagent itself in the solvent; this is then diluted to an exactly known volume. In either method, a highly purified chemical compound—called a primary standard—is required as the reference material. The process whereby the concentration of a standard solution is determined by titration of a primary standard is called a standardization.

The goal of every titration is the addition of standard solution in such amount as to be chemically equivalent to the substance with which it reacts. This condition is achieved at the equivalence point. For example, the equivalence point in the titration of sodium chloride with silver nitrate is attained when exactly one formula weight of silver ion has been introduced for each formula weight of chloride ion present in the sample. In the titration of sulfuric acid with sodium hydroxide, the equivalence point occurs when two formula weights of the latter have been introduced for each formula weights of the former.

The equivalence point in a titration is a theoretical concept. In actual fact we can only estimate its position by observing physical changes associated with it in the solution. The point in titration where such changes manifest themselves is called the end point. it is to be hoped that the volume difference between the end point and the equivalence point will be small. Differences do arise, however, owing to inadequacies in the physical changes and our ability to observe them. This results in an analytical error called a titration error.

One of the common methods of end-point detection employed in volumetric analysis involves the use of supplementary chemical compounds that exhibit changes in color as a result of concentration changes occuring near the equivalence point. Such substances are called indicators.

12. pipets, burets, volumetric flasks

All pipets are designed for the transfer of known volumes of liquid from one container to another. Some deliver a simple, fixed volume, these are called volumetric, or transfer pipets. Others, known as measuring pipets, are calibrated in convenient units so that any volume up to the maximum capacity can be delivered. Burets, like measuring pipets, enable the analyst to deliver any volume up to the maximum capacity. The precision attainable with a buret is appreciably better than that with a pipet. In general, a buret consists of a calibrated tube containing the liquid and a valve arrangement by which flow from a tip can be controlled. Volumetric flasks are available with capacities ranging from 5 mL to 5 liters, and are usually calibrated to contain the specified volume when filled to the line etched on the neck.

13. desiccator, crucible

Desiccators provide a measure of protection from the up-take of moisture. Simple crucibles serve as containers only. Of the two varieties avaliable the more common comprises those whose weight remains constant within the limits of experimental error while in use. The other crucibles serve as containers for the high-temperature fusion of difficultly soluable samples. These are appreciably attacked by their contents; the fused mass will thus hold contaminants derived from the crucible. Filtering crucible may be subdivided into two groups. There are those in which the filtering medium is an integral part of the crucible. In addition, there is the Gooch crucible, in which a filer mat (usually, but not always, asbestos, a mineral) is supported upon a perforated bottom.

14. precipitaion titration, complexometric titration, neutralization titration, oxidation-reduction titration

Volumetric methods based on the formation of a slightly soluble product (complexes) are called precipitation (complexometric) titrations. All end points employed in neutralization titration are based upon the sharp changes in pH that occur near the equivalence points; the range over which they occur varies considerably. The successful application of an oxidation-reduction reaction to volumetric analysis requires, among other things, the means for detecting the point of chemical equivalence in the reaction.

15. electroanalytical chemistry, electrogravimetric method

The field of electroanalytical chemistry encompasses a wide variety of techniques based upon the various phenomena occuring within an electrochemical cell. Many electroanalytical methods can be classified as gravimetric or as volumetric procedures. Thus, the use of an electric current to deposit a substance in a form suitable for weighing can be considered a variety of gravimetric analysis. Similarly, the use of any of several electrical properties will serve to establish the end point in an ordinary volumetric titration. In addition to these, other electroanalytical techniques measure some colligative electrical property that can be related by calibration to the concentration of the species being determined.

Electrolytic precipitation provides a simple method of isolating a number of elements from aqueous solutions. For our purposes most electrogravimetric procedures can be grouped in one of two basic categories. In the first the applied cell potential is continuously adjusted to maintain the potential of the electrode at which deposition occurs at some constant and predetermined value. In the other, no attempt is made to control the potential of the working electrode; instead, the potential applied is simply one that maintains a convenient current flow through the cell.

16. potentiometric titration

It is well known that the potential of a metallic conductor immersed in an electrolyte solution may be sensitive to the concentration of one or more of the components of that solution. An obvious analytical application of this phenomenon is its employment in the measurement of concentration. This may be done as follows: The progress of a titration is followed by measurement of the potential of an electrode immersed in the solution being titrated. This electrode must clearly be sensitive to the concentration of one of the participants of the analytical reaction; when this condition exists, an end point can be located by means of the behavior of the electrode. This procedure is called potentiometric titration.

17. conductometric titration

The conduction of an electric current through an electrolyte solution involves a migration of positively charged species toward the cathode and negatively charged ones toward the anode. All of the charged particles contribute to the conduction process; the contribution of any given species, however, is governed by its relative concentration and the inherent mobility of its individuals. The principal applications of direct conductance measurements include the analysis of binary-electrolyte mixtures and the determination of the total electrolyte concentration of a solution. A more important analytical application of conductance measurements involve their use to signal the end point in a titration. The main advantage of the conductometric titration is its applicability to very dilute solutions and to systems with relatively incomplete reactions.

18. coulometric titration

In coulometric analyses, the quantity of electricity is employed to determine the amount of a chemical substance in solution. One general technique makes use of a constant current that is passed until an indicator signals completion of the reaction. The quantity of electricity required to attain the end point is then calculated from the current and the time of its passage. This method is frequently called a coulometric titration.

19. methods based on absorption of radiation

The selective absorption of electromagnetic radiation as it passes through a solution causes the emerging beam to differ from the incident one. In the case of visible radiation, this difference is frequently obvious to the naked eye. For example, a white light viewed through a cupric sulfate solution appears blue because the cupric ions interact with and absorb the red components of the beam while transmitting completely the blue portions of the radiation. The application of this type of absorption phenomenon to the qualitative identification of substances is certainly familiar to most of us. Of equal importance is the employment of light absorption for the quantitative measurement of chemical systems.

20. methods based on emission of radiation

Upon being excited by a suitable energy source, the elements contained in a sample will emit visible and ultraviolet radiation. The wavelengths emitted are characteristic of the elements present; the intensity of the radiation is in part dependent upon the concentration. Thus, both qualitative and quantitative information can be gained through measurement of the wavelengths as well as the intensities of the radiations emitted by an excited sample. The analytical method based on this principle is called emission spectroscopy. In flame photometry, thermal excitation is accomplished by spraying a solution of a sample into the flame of a burner. The resulting spectra are generally less complex than those obtained by electrical excitation because the energies involved are considerably smaller.

21. mass spectrometer, calorimeter, barometer, manometer

In a mass spectrometer, a beam of positive ions is spread out according to the mass-to-charge ratio of the ions. Mass spectra are used to determine isotopic masses and abundances, molecular masses and structures, and to identify unknown compounds. Heats of combustion are measured in bomb calorimeter. For reactions that take place in solution, heat flow is measured in a calorimeter at atmospheric pressure. A barometer is an instrument that measures atmospheric pressure. A manometer is an instrument used to measure gas pressure in closed systems.

22. electromagnetic radiation and spectra, molecular spectroscopy

When electromagnetic radiation interacts with matter, atoms and molecules may absorb energy. Emission of this energy as radiation in various parts of the electromagnetic spectrum produces an emission spectrum. The sun and heated solids emit continuous spectra containing radiation of all wavelengths in a spectral region. Line spectra are produced when radiation is emitted only at specific wavelengths by excited electrons in atoms. Band spectra consisting of closely spaced lines are produced by excited molecules. Lines and bands may also be found in absorption spectra, produced when a continuum of radiation is passed through a sample that absorbs certain wavelengths. A spectrometer records the intensity and frequency of absorbed or emitted radiation.

Spectroscopy in different regions of the electromagnetic spectrum gives different kinds of information about the structure and geometry of molecules. Radiation at different wavelengths excites molecules in different ways, depending upon the match between the energy of the radiation and the energy of the various types of motion of the molecules and their atoms. Over the years spectroscopic techniques have become increasingly sophisticated, allowing the accumulation of more and more knowledge about molecules.

One major advance has been the coupling of computers to spectrometers, permitting the direct conversion of spectral data into whatever form is desired. Another major advance has been the advent of lasers as radiation sources in spectroscopy.

23. diffraction (X-rays, electrons, neutrons)

The regularly spaced atoms, ions, or molecules in a crystal can diffract beams of X-rays, electrons, and neutrons, providing information about the structure of crystals and of individual molecules. From the X-ray diffraction studies the distance between planes of atoms in the crystal can be calculated, and the dimensions of the unit cell found. The intensity of the diffracted radiation varies with the type and location of atoms or ions in the unit cell, providing information about the structure of molecules in molecular crystals or the arrangement of ions in ionic crystals. Low-energy electron diffraction (LEED) can be used to study atomic arrangements and electron densities at surfaces. Beams of neutrons are scattered solely by atomic nuclei and not by orbital electrons. This technique also yields information about the magnetic structure of materials.

24. chromatography

Chromatography is the distribution of a solute between a stationary and a mobile phase. It allows substances to be separated on the basis of their relative affinities for the two phases. Generally the stationary phase is a solid or a liquid, while the moving phase is a liquid or a gas. There are many versions of this technique. All depend on the fact that because one phase is in motion, equilibrium can never be reached, and so dissolved material moves continuously from the stationary into the mobile phase.

25. nuclear magnetic resonance

Nuclear magnetic resonance (NMR) spectroscopy is the study of the structure of molecules as revealed by the absorption of radio frequency radiation by nuclei. When the nuclei being studied are protons, the technique is referred to as proton magnetic resonance (PMR). Some of the other nuclei that have net spin and can also be studied by nuclear magnetic resonance are deuterium (^2H), boron (^{11}B), carbon (^{13}C), and oxygen (^{17}O).

26. infrared and ultraviolet spectroscopy

Infrared spectra are produced when radiation in the infrared region of the spectrum (2.5 μm to 25 μm) is absorbed by the atoms in a molecule, increasing their energy of motion with respect to each other. The absorption peaks occur at characteristic wavelengths which reflect the stretching, bending, deformation, or other distortions of various bonds in the molecule. An IR spectrum can often be used as a "fingerprint" to identify an unknown compound, or at least to determine what functional groups it contains. Absorption in the near-ultraviolet region of the spectrum (200~400 nm) requires the presence of the π electrons, that is, of multiple bonds. UV spectra, although simpler than IR spectra, can be used for structural analysis and compound identification, and also in quantitative analysis for determining the concentration of substances in solution.

27. laser spectroscopy, electron spectroscopy for chemical analysis (ESCA)

Laser Raman microanalysis (LRMA) refers to any technique in which a specimen is bombarded with a finely focused laser beam (diameter less than 10 μm) in the ultraviolet or visible range and the intensity versus wavelength function of the Raman radiation is recorded yielding information about vibrational states of the excited substance and therefore also about functional groups and chemical bonding.

Laser micro emission spectroscopy (LMES) refers to any technique in which a specimen is bombarded with a finely focused laser beam (diameter less than 10 μm) in the ultraviolet or visible range under conditions of vaporization and thermal excitation of electronic states of sample material and in which the photon emission spectrum is observed.

Laser micro-mass spectrometry (LMMS) refers to any technique in which a specimen is bombarded with a finely focused laser beam (diameter less than 10 μm) in the ultraviolet or visible range under conditions of vaporization and ionization of sample material and in which the ions generated are recorded with a time-of-flight mass spectrometer.

Photoelectron spectroscopy (PES) is a spectroscopic technique which measures the kinetic energy of electrons emitted upon the ionization of a substance by high energy monochromatic photons. A photoelectron spectrum is a plot of the number of electrons emitted versus their kinetic energy. The spectrum consists of bands due to transitions from the ground state of an atom or molecular entity to the ground and excited states of the corresponding radical cation. Approximate interpretations are usually based on "Koopmans theorem" and yield orbital energies. PES and UPS (UV photoelectron spectroscopy) refer to the spectroscopy using vacuum ultraviolet sources, while ESCA (electron spectroscopy for chemical analysis) and XPS use X-ray sources.

第14章 高分子化学术语
Chapter 14　Polymer Chemical Terms

14.1　Polymers

Polymers, or **macromolecules**, are produced when large numbers of smaller molecules (**monomers**) are bonded together. A **thermoplastic polymer** (or "plastic") softens when heated and resolidifies when cooled. The majority of common synthetic polymers have molecular masses between 1×10^4 and 1×10^6 u and contain several hundred to several thousand monomers. The physical properties of a polymer are determined by such factors as the flexibility of the macromolecules, the sizes of the groups attached to the polymer chains, and the magnitude of intermolecular forces. Polymers may be linear or branched; if the branches of one polymer chain are covalently bonded to other chains, the polymer is said to be cross-linked. **Thermoset polymers** are crosslinked polymers that are permanently rigid; they do not melt when heated. A **homopolymer** is formed by polymerization of a single type of monomer, a **copolymer** by the polymerization of two or more different monomers. Most polymers are either amorphous or semicrystalline. At low temperatures an amorphous polymer is rigid and brittle. At the **glass transition temperature**, T_g, the flexibility of the polymer increases markedly. Rigid plastics are used at temperatures below T_g, flexible plastics at temperatures above T_g. To participate in polymerization, a molecule must be able to react at both ends. The principal types of polymerization reactions are **chain reaction polymerization** and **step reaction polymerization.**

14.2　Natural Polymers

The unique properties of living things are in many ways the result of the properties of polymers. We now know that many of the molecules of biochemistry are macromolecules. Two natural polymers—cellulose and rubber—were the first polymers to be put into commercial use. Each had to be structurally modified to achieve useful properties.

1. Cellulose

The repeating unit in natural cellulose is a ring structure with numerous hydroxyl groups which provide strong hydrogen bonding between the chains. Microcrystalline regions also help to hold the chains together. Natural cellulose is not thermoplastic. Before the hydrogen bonds are broken and the microcrystals melted—a requirement for plastic flow—the molecule undergoes thermal decomposition. By converting some of the hydroxyl groups to functional groups that do not form hydrogen bonds, it is possible to alter the properties of the natural polymer. The nitrate ester of cellulose was prepared and used commercially in lacquers and films in the 1800s, before the nature of polymers was understood. (Due to its high flammability, the use of cellulose nitrate in such applications is now obsolete.) Cellulose acetate and other cellulose esters are still used in fibers. (Garments made of such fibers are labeled "acetate.")

2. Natural Rubber

Polymers that have elasticity, like rubber, are referred to as **elastomers**. Such polymers are utilized above their glass transition temperatures. The polymer chains in an elastomer must be flexible and they must be joined by a moderate number of crosslinks. In the unextended state, the molecules of an elastomer are tangled together. When a force is applied, the polymer chains move into an extended and more ordered arrangement. The crosslinks are necessary to prevent the chains from slipping past each other when force is applied. When the force is released the chains return to their less ordered state, largely because of the concomitant increase in entropy.

Natural rubber is an unsaturated polymer in which the chain enters and leaves each covalent double bond in a *cis* orientation. (The *trans* isomer is not rubbery.) Note that there are no crosslinks between the polymer chains. Natural rubber becomes soft and sticky when heated and tends to become permanently deformed when stretched. However, when natural rubber is heated with sulfur and a catalyst, sulfur cross-links are formed between the chains. This process is called **vulcanization**; it was discovered accidentally by Charles Goodyear in 1839. Introduction of a moderate number of sulfur cross-links converts natural rubber into a useful elastomer. A large number of cross-links convert it into hard rubber, or ebonite, which is not an elastomer.

14.3 Polymer Nomenclature

Considering that a simple compound like C_2H_5OH is variously known as ethanol, ethyl alcohol, grain alcohol, or simply alcohol, it is not too surprising that the vastly more complicated polymer molecules are also often known by a variety of different names. The International Union of Pure and Applied Chemistry (IUPAC) has recommended a system of nomenclature based on the structure of the monomer or repeat unit. A semisystematic set of trivial names is also in widespread usage; these latter names seem even more resistant to replacement than is the case with low molecular weight compounds. Synthetic polymers of commercial importance are often widely known by trade names which seem to have more to do with marketing considerations than with scientific communication. Polymers of biological origin are often described in terms of some aspect of their function, preparation, or characterization.

If a polymer is formed from a single monomer, as in addition and ring-opening polymerization, it is named by attaching the prefix **poly** to the name of the monomer. In the IUPAC system, the monomer is named according to the IUPAC recommendations for organic chemistry, and the name of the monomer is set off from the prefix by enclosing the former in parentheses. Variations of this basic system often substitute a common name for the IUPAC name in designating the monomer. Whether or not parentheses are used in the latter case is influenced by the complexity of the monomer name; they become more important as the number of words in the monomer name increases. The polymer $-(CH_2-CHCl)_n-$ is called poly (1-chloroethylene) according to the IUPAC system; it is more commonly called poly (vinyl chloride) or polyvinylchloride. Acronyms are not particularly helpful but are an almost irresistible aspect of polymer terminology, as evidenced by the initials PVC, which are so widely used to describe the polymer just named. The trio of names—poly (1-hydroxyethylene), poly (vinyl alcohol), and polyvinyl alcohol—emphasizes that the polymer need not actually be formed from the reaction of the monomer named; this polymer is formed by the hydrolysis of poly (1-acetoxyethylene), otherwise known as poly (vinyl acetate). These name alternatives are used in naming polymers formed by ring-opening reactions; for example, poly (6-aminohexanoic acid), poly (6-aminocaproic acid) and poly (ε-caprolactam) are all more or less acceptable names for the same polymer.

Those polymers which are the condensation product of two different monomers are named by applying the preceding rules to the repeat unit. For example, the polyester formed by the condensation of ethylene glycol and terephthalic acid is called poly (oxyethylene oxyterephthaloyl) according to the IUPAC system, as well as poly (ethylene terephthalate) or polyethylene terephthalate.

The polyamides poly (hexamethylene sebacamide) and poly (hexamethylene adipamide) are also widely known as nylon-6,10 and nylon-6,6, respectively. The numbers following the word *nylon* indicate the number of carbon atoms in the diamine and dicarboxylic acid, in that order. On the basis of this same system, poly (ε-caprolactam) is also known as nylon-6.

Most of the polymers to be given below are listed with more than one name. Also listed are some of the patented trade names by which these substances—or materials which are mostly of the indicated structure—are sold commercially.

Some commercially important cross-linked polymers go virtually without names. These are heavily and randomly cross-linked polymers which are insoluble and infusible and therefore widely used in the manufacture of such molded items as automobile and household appliance parts. These materials are called resins and, at best, are named by specifying the monomers which go into their production. Often even this information is sketchy. Examples of this situation are provided by phenol-formaldehyde and urea-formaldehyde resins.

e.g. 1 polyethylene; polystyrene; poly (vinyl chloride), "vinyl"; polyacrylonitrile, "acrylic"; poly(vinylidene chloride); poly(methyl methacrylate), plexiglass, lucite; polyisobutylene; polytetrafluoroethylene, teflon; poly (ethylene oxide), carbowax; poly (ε-caprolactam), nylon-6.

e.g. 2 polyester: poly(ethylene terephthalate), terylene, dacron, mylar; polyamide: poly(hexamethylene adipamide), nylon-6,6; polyurethane: poly(tetramethylene hexamethylene urethane), spandex, perlon U; polycarbonate: poly(4,4-isopropylidenediphyenylene carbonate), bisphenol A polycarbonate, Lexan; inorganic polymer: poly(dimethyl siloxane).

e.g. 3 common name—IUPAC name: α-(tribromomethyl)-ω-chloro-poly (1, 4-phenylene methylene); polyethylene—poly(methylene); polyacetylene—poly(vinylene); polypropene—poly (propylene); polyisobutylene—poly (1, 1-dimethyl-ethylene); polybutadiene—poly(1-butenylene); polyisoprene—poly(1-methyl-1-butenylene); polystyrene—poly (1-phenylethylene); poly (vinyl chloride)—poly (1-chloroethylene); polytetrafluoroethylene—poly(difluoromethylene); poly(methyl methacrylate)—poly(1-methoxycarbonyl-1-methyl-ethylene); poly(2,6-dimethylphenylene oxide)—poly(oxy-2,6-dimethyl-1,4-phenylene); poly(ethylene oxide)—poly(oxyethylene); polycyclopentene—poly(1,2-cyclopentylene) or polycyclopentane-1,2-ylene.

Basic Terms high (molecular weight) polymer, macromolecules (giant molecules), polymer, oligomer (having 2~10 monomer units only, low molecular weight polymer), synthetic plastics, synthetic rubber, synthetic fiber, monomer, regular polymer, irregular polymer, constitutional unit, constitutional repeating unit, configurational unit, degree of polymerization, tacticity, isotactic polymer, syndiotactic polymer, atactic polymer, stereoregular polymer, homopolymer, copolymer, alternating copolymer, random copolymer, block, block copolymer, graft copolymer, ionomer, interpenetrating polymer network (IPN).

14.4 Historical Introduction

The hypothesis that high polymers are composed of covalent structures many times greater in extent than those occurring in simple compounds, and that this feature alone accounts for the characteristic properties which set them apart from other forms of matter, is in large measure responsible for the rapid advances in the chemistry and physics of these substances witnessed in recent years (as of 1950s). This elementary concept did not gain widespread acceptance before 1930, and vestiges of contrary views remained for more than a decade thereafter. The older belief that colloidal aggregates, formed from smaller molecules through the action of intermolecular forces of mysterious origin, are responsible for the properties peculiar to high polymers is repudiated by this hypothesis. Such characteristic properties as high viscosity, long-range elasticity, and high strength are direct consequences of the size and constitution of the covalent structures of high polymers. Intermolecular forces profoundly influence the properties of high polymers, just as they do those of monomeric compounds, but they are not primarily responsible for the characteristics which distinguish polymers from their molecularly simple analogs. As a corollary of the prevailing viewpoint, the forces binding atoms of high polymers to one another, i. e., the covalent bonds, may be considered entirely equivalent to those which occur in analogous monomeric substances; intermolecular forces likewise are of a similar nature.

The implications of the foregoing concept have profoundly influenced modern trends in polymer research. If polymers owe their differences from other compounds to the extent and arrangement of their "primary valence" structures, the problem of understanding them is twofold. It is necessary in the first place to provide appropriate means, both experimental and theoretical, for elucidating their macromolecular structures and for subjecting them to quantitative characterization. Secondly, suitable relationships must be established to express the dependence of physical and chemical properties on the structures so evaluated. It may appear incredible that significant investigations from this obvious point of view were not undertaken in so manifestly important a field before about 1930, and that noteworthy advances have occurred principally since 1940. The reasons for the delay in the evolution of a rational approach to the study of high polymers would be difficult to explain adequately in few words. An insight into them and a perspective with respect to the course of more recent investigations may be gained, however, through a survey of the circumstances leading up to the eventual acceptance in the early 1930s of the primary valence, or "molecular" viewpoint concerning the constitution of polymeric substances.

The earliest reported studies of polymeric substances which may be regard as significant from the point of view expressed above were carried out by two essentially independent groups of investigators. On the one hand, there were those concerned with the physical and chemical constitution of the natural polymers—starch, plant fibrous material (cellulose), proteins, and rubber. The other group consisted of the synthetic organic chemists of the latter half of the nineteenth century who, though not primarily interested in polymeric substances, inadvertently came upon synthetic polymers incidental to the pursuit of other objectives. Neither group of investigators appears to have been aware of the significance with reference to the naturally occurring polymers of these occasionally reported syntheses of polymeric products. The conclusions which were reached from results obtained in each of these two fields will be examined below.

(1) Early investigations on naturally occurring polymers

(2) Early encounters with condensation polymers

(3) Vinyl polymers

(4) Rise of the macromolecular hypothesis

In an important paper published in 1920 Staudinger deplored the prevailing tendency to formulate polymeric substances as association compounds held together by "partial valences". He specifically proposed the chain formulas $—CH_2—CH(C_6H_5)—CH_2—$

$CH(C_6H_5)-$, etc. and $-CH_2-O-CH_2-O-CH_2-O-$, etc. for polystyrene and polyoxymethylene (paraformaldehyde), which are the ones accepted at the present time. He also advocated the long chain formula for rubber. The colloidal properties of these substances were attributed entirely to the sizes of their primary valence molecules, which he guessed might contain of the order of a hundred units. Staudinger disposed of the end group problem with the utmost facility by simply suggesting that no end groups are needed to saturate terminal valences of the long chains. At that time it seemed not implausible that free radicals at the ends of very long chains would be unreactive owing to the size of the molecule. He called attention also to the frequent erroneous assignment of relatively simple cyclic structures to substances which actually are chain polymers, mentioning as examples dimethylmalonic anhydride, adipic anhydride, and glycolide.

Staudinger relentlessly championed the molecular, or primary valence, viewpoint in the years which followed. He supported his original contentions with the observation that hydrogenation of rubber, as well as its conversion to other derivatives, does not destroy its colloidal properties. In contrast to association colloids, high polymers (or **macromolecules** as he chose to call them) exhibit colloidal properties in all solvents in which they dissolve. Polyoxymethylenes were extensively investigated, leaving no real basis for doubt as to their linear polymeric nature.

The views of Staudinger were not widely accepted at once, most investigators tenaciously adhering to the finite certainty offered by ring formulas in preference to the vagaries of chains of undefined length. (Duclaux declared in 1923 that it is useless to seek explanations for the properties of rubber on the basis of its molecular structure inasmuch as rubber should be regarded as a physical state rather as a chemical compound.) Molecular weight measurements which yielded moderately low values, now known to have been seriously in error, seemed to support this opposition. Ring formulas composed of one or several $C_6H_{10}O_5$ units were advocated for cellulose and starch. X-ray diffraction indicated unit cells for crystalline rubber and cellulose similar in size to those of simple substances, from which it was argued that the molecules must likewise be small. The fallacy in the assumption that the molecule could be no larger than the unit cell had been pointed out earlier, but it remained for Sponsler and Dore to show in 1926 that the results of X-ray diffraction by cellulose fibers are consistent with a chain formula composed of an indefinitely large number of units. The structural units occupy a role analogous to that of the molecule of a monomeric substance in its unit cell; the cellulose molecule continues from one unit cell to the next through the crystal lattice. This interpretation nullified one of the final arguments mustered in support of the associated ring theory; it was soon extended to other linear polymers showing characteristic X-ray fiber patterns.

Meyer and Mark, who were among the foremost of the early advocates of long chain, covalent structures for polymers, retained the association hypothesis to the extent that they regarded the crystallites to be discrete units or "micelles" formed by aggregation of polymer molecules. They proceeded in 1928 to estimate the size of the crystallite from the breadths of the X-ray diffraction spots. In this way, lengths of 50 to 150 units were deduced for the cellulose and rubber micelles. Micelle length was considered to be identical with molecular length, molecular weights of the order of 5000 being estimated in this manner. The much higher values, 150 000 to 400 000, which they obtained from careful osmotic pressure measurements on dilute rubber solutions were attributed to solvation, and later were cited as evidence that the micelle observed by X-rays persists in solution.

Staudinger's opposing view that the size of the crystallite bears no relationship to the size of the polymer molecule has been largely substantiated. Thus, neither the dimensions of the unit cell nor those of the entire crystallite are related (directly at least) to the polymer chain length. A polymer molecule may pass through the many unit cells reaching from one end of a crystallite to the other, then meander through an amorphous region and into another crystallite, etc. Extending the covalent structure idea one step further, Staudinger in 1929 differentiated linear from nonlinear, or **network**, polymers. He attributed the characteristic infusibility and insolubility of the latter to the formation of network structures of great extent. Meyer and Mark, a year earlier, had proposed that the properties of vulcanized rubber are to be accounted for by the formation of covalent crosslinkages, a view subsequently confirmed.

In 1929 Carothers embarked on a series of brilliant investigations which were singularly successful in establishing the molecular viewpoint and in dispelling the attitude of mysticism then prevailing in the field. The object of these researches, clearly expressed at the outset, was to prepare polymeric molecules of definitive structures through the use of established reactions of organic chemistry, and further to investigate how the properties of these substances depend on constitution. The major contributions of Carothers and his collaborators lay in the field of condensation polymers—the polyesters, the polyamides, and so forth. In a sense, Carothers extended the synthetic approach emphasized by Emil Fischer (and intimated much earlier in the work of Lourenço), discarding, however, the unnecessary and severely encumbering insistence on pure chemical individuals under which Fischer and his colleagues labored.

Scarcely had the covalent chain concept of the structure of high polymers found root when theoretical chemists began to invade the field. In 1930 Kuhn published the first application of the methods of statistics to a polymer problem; he derived formulas expressing the molecular weight distribution in degraded cellulose on the assumption that splitting of interunit bonds occurs at random.

The statistical approach has since assumed a dominant role in the treatment of the constitution, reactions, and physical properties of polymeric substances. The complexities of high polymers are far too great for a direct mechanistic deduction of properties from the detailed structures of the constituent molecules; even the constitution of polymers often is too complex for an exact description. The very complexities which make the task of rational interpretation of polymers and their properties appear formidable actually provide an ideal situation for the application of statistical procedures.

The average size and shape of a long chain molecule endowed with the ability to assume all sorts of configurations through rotations about its valence bonds have held a fascination for theoretically minded chemists and physicists for many years. In 1934, Guth and Mark, and Kun independently discussed the problem, arriving at similar solutions. These theoretical investigations furnished the background for an attack on such problems as the high viscosities exhibited by dilute solutions of high polymers, double refraction of flow, and rubber elasticity.

The quantitative treatment of polymer constitution and behavior, which is a necessary counterpart of the theoretical approach, could not take place until methods for quantitatively characterizing polymer structures had been established. The various properties of polymers which depend on molecular weight obviously could not be satisfactorily elucidated prior to the establishment of means for assigning correct molecular weight values. Associated therewith is the related problem of characterizing molecular weight distributions. Staudinger deserves credit for emphasizing that the molecular weight of a linear polymer may be related directly to the viscosity of its dilute solutions. Solution viscosity is a readily measured quantity, and such measurements are widely used technically for this reason. The relationship between polymer solution viscosity and the molecular weight of the polymer is therefore of considerable utility for both pure and applied investigations. Staudinger erroneously concluded, however, that a direct proportionality exists between the quantity now called the **intrinsic viscosity** and the molecular weight. Widespread use of Staudinger's formula led to the assignment of molecular weights which were generally too low, sometimes by factors of ten or more. This situation was not fully rectified until the mid-1940s. At present a rather impressive number of different polymer series have been subjected to quantitative molecular weight measurements, and reliable relationships between solution viscosity and molecular weight of the polymer have been established in many cases.

Not until the 1940s did suitable methods, both experimental and theoretical, become available for reducing the constitution of polymers, including the nonlinear, network-forming types, to tractable quantitative terms. Since such means are a prerequisite to the quantitative treatment of polymer properties in relation to constitution, advances in this direction necessarily were delayed.

14.5 Recycling Synthetic Polymers

You finish off the last of the milk. What are you going to do with the empty bottle? If you toss it into the trash, it will almost certainly go into a landfill, taking up space and serving no useful purpose. But if you put it in the recycle bin, it's likely to be melted down to produce something new.

Between 50 and 60 billion pounds of synthetic polymers are manufactured each year in the United States—over 200 pounds per person. A large percentage of these polymers are tossed into our landfills after use. This represents a serious waste of precious raw materials (the petroleum products from which synthetic polymers are made) and exacerbates concern that the landfills are quickly filling up. These factors give the recycling of polymers a high priority.

Some synthetic polymers can be recycled and some cannot. So-called thermoplastic polymers, usually composed of linear or only slightly branched molecules, can be heated and formed and then reheated and reformed. On the other hand, thermosetting polymers, which consist of molecules with extensive three-dimensional cross-linking decompose when heated, so they cannot be reheated and reformed. In general, thermosetting polymers cannot be recycled.

In 1988 the Plastic Bottling Institute suggested a system in which numbers embossed on objects made of polymers tell the recycling companies what type of polymers was used in the object's construction (Table 14.1). This numbering system facilitates the collection, sorting, and reprocessing of the polymers that can be recycled.

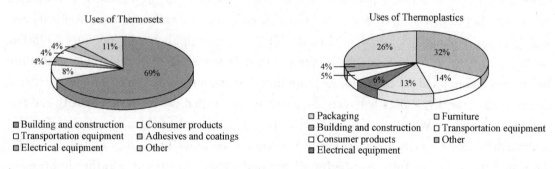

Table 14.1 Recyclable Thermoplastics

Symbol and abbreviation	Name of polymer	Examples of uses for virgin polymer	Examples of uses for recycled polymer
♲ 1 PET	polyethylene terephthalate	beverage containers, boil-in food pouches, processed meat packages	detergent bottles, carpet fibers, fleece jackets
♲ 2 HDPE	high-density polyethylene	milk bottles, detergent bottles, plastic buckets, mixing bowls, oil bottles, toys, plastic bags	compost bins, detergent bottles, crates, agricultural pipes, curbside recycling bins

Symbol and abbreviation	Name of polymer	Examples of uses for virgin polymer	Examples of uses for recycled polymer
3 PVC	poly (vinyl chloride)	food wrap, vegetable oil bottles, blister packaging, plastic pipes	detergent bottles, tiles, plumbing pipe fittings
4 LDPE	low-density polyethylene	shrink wrap, plastic sandwich bags, garment bags, squeeze bottles	films for industry and general packaging
5 PP	polypropylene	margarine and yogurt containers, grocery bags, caps for containers, carpet fiber, food wrap, plastic chairs, luggage	compost bins, curbside recycling bins
6 PS	polystyrene	plastic utensils, clothes hangers, foam cups and plates	coat hangers, office accessories, video/CD boxes
7 Other	includes nylon	other	—

第15章 生物化学术语
Chapter 15　Biochemical Terms

15.1　Biochemistry

Most biomolecules are organic polymers. The structures and function of biological molecules are intimately related. Cellular membranes and other structural elements are composed chiefly of hydrophobic substances. Most of the substances that take part in the chemical reactions of organisms are hydrophilic. Carbohydrates are simple sugars (monosaccharides), combinations of two sugars (disaccharides), or polymers of many sugars (polysaccharides). Simple sugars are polyhydroxy aldehydes (aldoses) or ketones (ketoses). Glucose, like many monosaccharides, exists chiefly in a ring configuration in solution. Polysaccharides are used by organism as storage forms for sugars (starch, glycogen) or to provide structural support (cellulose). **Lipids** are all nonpolymeric, hydrophobic substances. Among the lipids are glycerides (fats and oils), esters of fatty acids, and the alcohol glycerol. Phospholipids play important roles in cellular membranes because they have polar, hydrophilic heads and nonpolar, hydrophobic tails. **Proteins** consist of one or more long, unbranched polymer chains of **amino acids** linked by **peptide bonds** (**polypeptides**) [**primary** (1°) **structure**]. Fibrous proteins are generally hydrophobic, structural materials in which the polypeptide chains are relatively extended and cross-linked [**secondary** (2°) **structure**]. Globular proteins are more compact, elaborately folded molecules, usually with hydrophilic exteriors that allow them to function in aqueous solution [**tertiary** (3°) **or quaternary** (4°) **structure**]. Often they contain more than one polypeptide chain. **Enzymes**, the catalysts of living systems, are large protein molecules. Enzyme molecules possess a groove called the active site, where substrate (the molecule or molecules with which an enzyme interacts) molecules can be bound. An enzyme brings substrate molecules together in proper orientation for reaction, weakens bonds in the substrate(s), and may take a temporary part in the reaction mechanism.

Nucleic acids are polymers of **nucleotides**, each of which consists of a five-carbon sugar (ribose or deoxyribose), a nitrogen-containing base, and a phosphate group. **DNA** (**deoxyribonucleic acid**) carries the genetic "blueprints" that specify what proteins an organism can make, while **RNA** (**ribonucleic acid**) is involved in "reading" the blueprints and guiding the synthesis of the proteins.

第 15 章 生物化学术语(Biochemical Terms)

The breakdown of large molecules, which generally yields energy, is called **catabolism**; the synthesis of large molecules from simpler components, which generally requires energy, is called **anabolism**. Together these two processes make up the organism's **metabolism**. Organisms use the energy released by catabolic reactions to make **ATP** (**adenosine triphosphate**) from **ADP** (**adenosine diphosphate**) and phosphate. Enzymes couple the reverse of this reaction—the hydrolysis of ATP—to anabolic reactions so that the energy provided by the former can drive the latter.

1. Name, Abbreviation, and R Group for Some Common Amino Acids

The backbone structure: $H_2N—CHR—COOH$

Name	Abbreviation	R
Alanine	Ala	$—CH_3$
Arginine	Arg	$—CH_2CH_2CH_2NHC(NH_2)=NH$
Aspartic acid	Asp	$—CH_2COOH$
Cysteine	Cys	$—CH_2SH$
Glutamic acid	Glu	$—CH_2CH_2COOH$
Glycine	Gly	$—H$
Isoleucine	Ile	$—CH(CH_3)CH_2CH_3$
Leucine	Leu	$—CH_2CH(CH_3)_2$
Lysine	Lys	$—CH_2CH_2CH_2CH_2NH_2$
Methionine	Met	$—CH_2CH_2SCH_3$
Phenylalanine	Phe	$—CH_2C_6H_5$
Serine	Ser	$—CH_2OH$
Threonine	Thr	$—CH(OH)CH_3$
Tyrosine	Tyr	$—CH_2C_6H_4OH(p-)$
Valine	Val	$—CH(CH_3)_2$

2. Three Levels of Structure in Protein Molecules or Biopolymers

(1) Primary structure refers to the sequence of amino acids in the polyamide chain. (Details of the microstructure of a chain is a description of the primary structure.)

(2) Secondary structure refers to the shape of the molecule as a whole, particularly to those aspects of structure which are stabilized by intermolecular hydrogen bonds. (The overall shape assumed by an individual molecule as a result of the rotation around individual bonds is the secondary structure.)

(3) Tertiary structure also refers to the overall shape of a molecule, especially to structures stabilized by disulfide bridges (cystine) formed by the oxidation of cysteine mercapto groups. (Structures that are locked in by chemical cross-links are tertiary structures.)

3. Examples of the Effects and Modification of the Higher-Order Levels of Structure in Proteins are Found in the Following Systems

(1) Collagen is the protein of connective tissues and skin. In living organisms, the

molecules are wound around one another to form a three-strand helix stabilized by hydrogen bonding. when boiled in water, the collagen dissolves and forms gelatin, apparently establishing a new hydrogen bond equilibrium with the solvent. This last solution sets up to form the familiar gel when cooled, a result of shifting the hydrogen bond equilibrium.

(2) Keratin is the protein of hair and wool. These proteins are insoluble because of the disulfide cross-linking between cystine units. Permanent waving of hair involves the rupture of these bonds, reshaping of hair fibers, and the reformation of cross-links which hold the chains in the new positions relative to each other. Such cross-linked networks are restored to their original shape when subjected to distorting forces.

(3) The globular proteins **albumin** in eggs and **fibrinogen** in blood are converted to insoluble forms by modification of their higher-order structure. The process is called denaturation and occurs, in the system mentioned, with the cookings of eggs and the clotting of blood.

Size and Complexity of Biopolymers

Class	Specific example	Typical mol wt
Oligomers	Actinomycin D (2 nm sphere)	$10^3 \sim 10^4$
Small proteins	Chymotrypsin (4 nm sphere)	$10^4 \sim 10^5$
Nucleic acids	tRNA (10 nm rod)	$10^4 \sim 10^5$
Large proteins	Aspartate transcarbamylase (7 nm sphere)	$10^5 \sim 10^7$
Small assemblies	Ribosome (20 nm sphere)	$10^5 \sim 10^7$
Large assemblies	Membrane, viruses (100 nm sphere)	$10^7 \sim 10^{12}$
Intact DNA	*E. coli* DNA (0.1 cm rod)	$10^7 \sim 10^{12}$

15.2 Levels of Structures in Biological Macromolecules

The molecular weights of many biological systems range from 10^3 to 10^{12} daltons. In many cases, such a system is composed of many covalent species rather than only one. The whole system is called a single molecule when the parts are present in a well-defined stoichiometry and when there is little tendency for them to dissociate spontaneously under physiological conditions.

These biomolecules can be divided into seven classes of progressively increasing size and complexity as shown before. The first three classes contain molecules for which structural information is potentially available at atomic resolution. It is meaningful and useful to frame questions about the structure and function of such molecules at the level of the location or behavior of individual atoms or individual **residues** (recognizable groups of atoms).

The last four classes represent systems of such complexity that, even if atomic resolution structures were available, such information would be extremely cumbersome to use. For example, a 10^6 dalton structure would consist of about 10^5 atoms. Describing such a structure by naming each atomic type and its spatial position would be analogous to describing a city by reciting the names and addresses of each of its inhabitants. A more manageable level of description involves naming individual subunits or regions, and then concentrating on individual residues or atoms only when their function is particularly interesting. This is akin to breaking the city down into neighborhoods or zones and keeping track of only a few significant individuals, such as mayor.

Ambiguities can easily arise when terms like subunit, residue, molecule, region, and moiety are used to describe biopolymers with any kind of generality. There is a hierarchy of structural units and structural features. Because the patterns of organization of proteins, nucleic acids, and complex noncovalent aggregates such as cell surfaces are somewhat different, no single nomenclature is optimal for describing all three types of structure. Furthermore, there is no universal agreement even on an exact terminology for proteins alone. Here, we shall introduce what we consider to be the most convenient way of classifying levels of macromolecular structure. Keep in mind that the following definitions cannot be viewed inflexibly because many biological structures have features that do not fall cleanly into one category or another.

The individual monomeric residues that form proteins and nucleic acids are amino acids and nucleotides, respectively. These generally have known atomic structure. That is, one can define the identity of every atom, the average location of its nucleus in space, where, on the average, the electron density is concentrated. Such structures are not rigid or inflexible, but molecular vibrations and internal rotations about bonds usually do not distort the picture too badly, especially in ordered crystals. The configuration of a monomer residue (as used in the language of organic chemistry) is its atomic structure, including known stereochemistry of asymmetric centers. The conformation is a more complete description and includes what is known about preferred orientations of groups that are, in principle, capable of movement by internal rotation. The conformation is generally an average over energetically accessible atomic structures and is the most complete description possible of a molecule in solution.

1. Primary Structure is the Sequential Order of the Residues

Proteins and nucleic acids are unbranched polymers. The covalent chain structure of either of these can be written $R_1 R_2 \ldots R_i \ldots R_n$, where R_i gives the identity of the ith residue in the chain. A feature of proteins, nucleic acids, and most polysaccharides that immediately distinguishes them from many synthetic polymers is that the units are arranged in a head-to-tail fashion. Thus, the abbreviations $R_a R_b$ and $R_b R_a$ refer to molecules that have different covalent chemical structures. Also, the two ends of the polymer $R_1 \ldots R_n$ will have different chemical properties. To describe the covalent structure, the absolute **sequence** of residues must be written. There may be more than one polymer chain in the complete covalent structure and these, in turn, may contain intrachain or interchain covalent cross-links. The complete covalent structure is called the primary structure (1° structure). It must include a specification of the configuration of all asymmetric centers, both in the polymer chain backbone and on the side chains of each of the monomer residues.

The primary structure of hundreds of proteins and nucleic acids has been determined by elaborate chemical analysis. An up-to-date compilation of all known primary structures is published regularly. It would be extremely time-consuming to write a chemical formula for the entire primary structure. Therefore, it is customary to abbreviate primary structures by using an alphabetic code. The most common version of this, (shown below) uses a one-letter code for each nucleotide and a three-letter code for each amino acid. The terms sequence and primary structure are often used interchangeably and, in fact, they are identical for all single-chain non-crosslinked structures. Note that neither term implies knowledge of the conformation of any individual residues or the polymer backbone.

Type	Example	Abbreviation
Polystyrene	⇒ CH_2—CHPh—CH_2—CHPh—CH_2—CHPh ⇒	
Polyamide	⇒ NH—$(CH_2)_6$—NH—CO—$(CH_2)_4$—CO—NH—$(CH_2)_6$—NH—CO ⇒	
Polypeptide	⇒ NH—CH(CH_3)—CO—NH—CH(CH_2OH)—CO—NH—CH(H)—CO ⇒	—Ala—Ser—Gly—
Polysaccharide		—NAG—NAM—NAG
Polynucleotide		ApUpC

2. Secondary Structure Describes Helices of Residues

Many biopolymer chains can form locally ordered, three-dimensional structures. For individual, linear, head-to-tail polymers, with asymmetric monomer units, the only kind of symmetric three-dimensional ordered structure possible is a helix. This contains a screw axis of symmetry.

Other helices can be formed by bringing together two or more individual helical strands. Double and triple helices are fairly common in nature. In multiple-stranded helices, the individual covalent chains can run in parallel directions, but it is also possible to construct helices in which some strands run in opposite directions. The Watson-Crick DNA double-helical structure is a 10-fold (10 residues per turn) double helix with antiparallel strands. Polypeptide β-sheets are multistranded twofold helices in which some neighboring strands run parallel whereas others run antiparallel.

The secondary structure (2° structure) of a biopolymer is defined as an enumeration of the particular regions of the primary structure involved in any kind of helix. More generally, secondary structure is a list of all three-dimensional regions that have ordered, locally symmetric backbone structures.

3. Tertiary Structure is the Three-Dimensional Arrangement of Residues

The tertiary structure (3° structure) of a protein or nucleic acid is the complete three-dimensional structure of one effectively indivisible unit. For a protein, this unit is usually one single covalent species, whether it contains a single polypeptide or more than one linked by covalent crosslinks. For a nucleic acid, it is either a single covalent strand in the case of most RNAs, or two complementary double strands in the case of most DNAs. The tertiary structure includes a description not only of local symmetric structure (2° structure) but also of the spatial location of all residues insofar as is possible.

The term conformation is used synonymously with tertiary structure. This is because, if one knows the conformation (i.e, the angle of rotation around all single bonds in a polymer), one indeed also knows the tertiary structure, and vice versa. Problems at the tertiary structure level do not necessarily imply that one knows or is interested in the entire three-dimensional structure.

4. Quaternary Structure is the Arrangement of Subunits

The highest level of structure we shall consider is quaternary structure (4° structure). This is formed by the noncovalent association of independent tertiary structure units. The subunits of the quaternary structure may or may not be identical, and their arrangement in the quaternary structure may or may not be symmetrical. Example of a simple quaternary structure is provided by vertebrate hemoglobins. These consist of four subunits—two each of two types (α and β), each of which is a single polypeptide chain folded into a compact globular tertiary structure. Each chain contains a bound heme group.

A typical more complex quaternary structure is the *E. coli* ribosome. This contains three RNA molecules and approximately 55 different protein chains. All but one of the latter are present in single copies; one protein occurs in four copies. For structures such as the ribosome that are really a hierarchy of quaternary structure levels, it is sometimes convenient to use terms such as "quinternary structure" to define increasing stages of association. However, we shall avoid these terms.

The quaternary structure of a few multisubunit proteins is known fairly accurately from X-ray diffraction studies. These structures range from proteins as simple as superoxide dismutase (with two identical subunits) to molecules as complicated as aspartate transcarbamylase. A number of much more complex quaternary structures are known approximately through the use of electron microscopy, sometimes assisted by low-resolution X-ray measurements. Examples of these are the *E. coli* pyruvate dehydrogenase complex (a very large assembled protein) and tobacco mosaic virus. There is some information available about quaternary structure of even more complex systems—e. g., protein assemblies in striated muscle fibers, and protein substructures within rather complex viruses such as T2 and T4 bacteriophages.

15.3 DNA and Proteins: Which Came First in the Life's Chemical Evolution?

The prebiotic assembly of both amino acids and nucleotides into polymeric forms is possible in principle, if it's difficult in practice. Amino acid polymerization is not easy in a watery environment. The formation of a peptide bond between amino acids involves the expulsion of a molecule of water; but the bond can be split again by water (a process known as hydrolysis). In the presence of a lot of water, hydrolysis will be more favorable than peptide formation. It is conceivable that amino acids could have become linked up into polypeptides by being concentrated in evaporating pools or in hot, dry environments close to volcanoes; but experiments that attempt to mimic such processes give only tiny yields of polypeptides. The linking together of amino acids can be assisted, however, by certain kinds of reactive molecules, called condensing agents, which can "soak up" the water that is produced. The compound cyanamide, for example, can induce linkage of the amino acids glycine and leucine (Figure 15.1). Cyanamide can itself be formed from hydrogen, cyanide under plausibly prebiotic conditions.

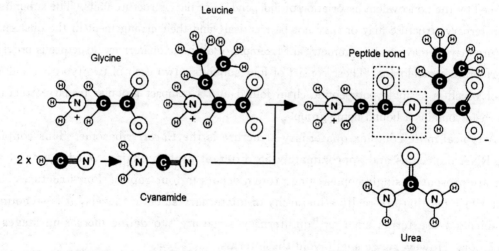

Figure 15.1 The amino acids glycine and leucine can be linked into a dipeptide using cyanamide as a "condensing agent"

John Oró, Cyril Ponnamperuma and others have shown that condensing agents such as cyanamide will also assist bases, sugars and phosphate to combine into nucleotide, while Leslie Orgel of the Salk Institute in California has found that metal ions such as zinc will help nucleotides to link up into oligonucleotides (short polymers containing a few nucleotides). Primed with polymerized nucleotides and polypeptides, we seem to be well on the road to a world containing the nucleic acids and proteins of primitive living organisms. Is our quest for a chemical origin of life near an end? Not a bit of it. We have just reached one of the most difficult hurdles of all.

So far we have been relying pretty much on chance. That is to say, we have supposed our raw materials to be created amidst a whole mess of other compounds from very crude reactions. And we have seen that it is possible for these compounds to link up in random fashion to give polymers. We now have to face the fact that life is anything but a random process; it is in fact just about the most impressive feat of molecular organization that we know of. Previously, it was suggested that a reasonable definition of life can be specified in terms of three functions: self-replication, self-repair and metabolism. Moreover, it is generally recognized that living systems must be bounded systems—they must have boundaries of some kind. All of these characteristics require a high degree of organization and cooperation at the molecular scale. Where does the organization come from? Surely not from the insensate, random world of prebiotic chemistry that has been described so far?

Well, it must, of course, unless we are going to shrug our shoulders and accede to divine intervention. Having got this far, I think that we would be advised to reserve that option for when we get really desperate!

The organization within an organism stems ultimately from its genetic makeup—its genome—which encodes the information for construction of the organism's molecular machinery. Most of this machinery is in the form of proteins, while the information itself is encoded on the nucleic acid DNA and is translated into proteins via the agency of RNA. Here we run up against one of the conundrums that defeated origin-of-life researchers for nearly three decades after the ground-breaking experiments of Miller and Urey. Assembling DNA-like oligonucleotides by the random joining of nucleotides has a vanishingly small chance of ever producing anything that resembles the blueprint for an organism, which is to say, for the protein enzymes that are essential to life. In just the same way, we can forget about the chance of making efficient enzymes from the random assembly of amino acids. Both DNA and proteins can be considered replete with meaningful information—they are preprogrammed for specific functions. But preprogrammed how?

The problem can be posed in another way. Proteins require DNA for their information, the protein plan being coded in the four-symbol DNA alphabet. But DNA—as distinct from random polynucleotide chains—cannot be created without the assistance of protein enzymes, which assist the assembly of new strands of nucleic acid on the templates of existing ones. If we can imagine a world with DNA but no proteins, we can see how the latter could be put together from the information encoded in the former. And if we imagine a world with protein enzymes but no DNA, it is conceivable that the enzymes could act together to synthesize nucleic acids from prebiotic nucleotides. But until we have one, we cannot have the other—and vice versa. The problem of the chicken and the egg, far from being a trival philosophical paradox, provides a singularly apt metaphor for the origin of life.

15.4 The RNA World

In the early 1980s, a possible way was discovered to break the protein-DNA circle. So far neglected in the discussion is the go-between in the process of translating DNA to proteins—the humble RNA molecule. But this functions as a truly replicating molecule. RNA is able both to store genetic information (recall that messenger RNA, or mRNA, carries the information encoded in genes) and to act as a template for the formation of proteins. In other words, it acts both as a carrier of the genetic plan of an organism (the genotype) and as a menas for the external expression of that genetic information (the phenotype), via the formation of proteins.

The idea that RNA might have been the first molecular replicator dates back to the 1960s, but such early speculations floundered on one crucial point: the replication of RNA seemed to suffer from the same limitation as that of DNA, in that it required help from enzymes. In the 1980s, however, the molecular biologists Sidney Altman and Thomas Cech discovered that this was not necessarily the case. They found that some RNA molecules can catalyze the assembly of other RNA molecules, thereby acting as "nonprotein enzymes". The implication is that such catalytic RNA molecules might be able to facilitate their own replication. Cech and Altman called these catalytic RNA molecular ribozymes.

The discovery of catalytic RNA, which won Altman and Cech the Nobel prize for chemistry in 1989, revitalized interest in the suggestion of a prebiotic world populated by replicating RNA molecules—not exactly living systems, but well on the way towards them. The biologist Walter Gilbertof Harvard University has christened this scenario the "RNA world". The first inhabitants of the RNA world would have been simple RNA-like oligonucleotides with some ability to catalyze their reproduction. Mutations in the nucleotide sequences would ocassionally arise from imperfact replication on the RNA template, and mutant forms that turned out to be better at replication would dominate over the others by Darwinian selection. These RNA replicators would become ever more efficient, and could eventually learn how to put together proteins, perhaps in a crude imitation of the codon-based translation process that goes on in our own cells. Some of these proteins might turn out to assist in RNA replication, making them primitive enzymes. RNA molecules that could produce their own enzymes would gain a tremendous evolutionary advantage over their less capable fellows, and every advance made in this ability would produce a new dominant strain. Only quite late in the day would DNA appear—a double-stranded version of RNA in which the base uracil had become replaced by thymine. As it is a more stable database for storing genetic information, DNA would have gradually taken over as the central components of a replicating system, subverting RNA to the role of an intermediary in protein synthesis.

The catalytic behavior of RNA provides one of the cruial links required for this picture to seem plausible, but it is by no means the only good reason for believing that life, in its earliest manifestation, arose from a RNA world. While DNA is generally a passive memory bank for genetic information, the various forms of RNA paly a very diverse and active part in the biological processes of the cell. In particular, they are central to processes that are thought to be of most ancient origin. Many coenzymes—molecules that assist enzymes in their various tasks—are based either on true RNA nucleotides or on related compounds, suggesting that RNA-like species may have had to exhibit a wide range of talents before proteins became the dominant biochemical catalysts.

Nevertheless, the scenario of the RNA world is not without its problems, not the least of which is the question of how RNA-like molecules evolved in the first place. We have seen that the basic components of nucleotides can be created from simple organic molecules, albeit with some considerable artifice. Nucleotides can be linked at random in the presence of condensing agents, but the oligonucleotides produced will represent gobbledigook in genetic terms, a far cry from information-laden RNA. It is likely that life did not really start with RNA at all, but with some simple, albeit similar, kind of molecule that also had some ability to replicate and carry crude genetic information. Perhaps sugars other than D-ribose were incorporated into these pre-RNA replicators, and truly RNA-like molecules were then gradually singled out from this jumble by virtue of their particular aptitude for reproducing themselves reliably. If this was so, what might these earliest of replicating molecules have looked like?

第16章 环境化学术语
Chapter 16　Environmental Chemical Terms

16.1　Prologue on Energy and Sustainability

The question of energy underlies virtually all environmental issues. The harnessing of energy for the manifold needs of industrial civilization has driven economic development, and access to affordable energy has been the key to a better life for people around the world. At the same time the environmental costs of human energy consumption are becoming ever more apparent: oil spills, the scarring of land by mining, air and water pollution, and the threat of global warming from the accumulation of carbon dioxide and other greenhouse gases. Inceasingly, maintaining an expanding supply of cheap energy seems to clash with concern for the environmental costs of such expansion.

Environmental discussion often revolve around sustainability, an evocative and much-debated idea. It arises from the perception that human activity is using up nature's resources at rates beyond the capacity of nature to restore them. Sustainability implies maintaining these resources for future generations.

The concept has many applications. Sustainability logging, for example, refers to extracting wood in ways that permit regeneration of forests. Sustainable agriculture would feed people without depleting the nutritive capacity of the soil or the biodiversity of natural habitats. A growing number of companies embrace sustainability by protecting the environment in ways that go beyond legal requirements. Examples include voluntarily curbing emissions, switching to environmentally benign materials (known as green chemistry), or refraining from using tropical hardwoods in furniture manufacture. The new field of Industrial Ecology contributes to sustainability by finding ways to minimize the consumption of material in industrial society.

However, there is wide argument about how to achieve sustainability. Does sustainable logging mean leaving the wildness undisturbed and harvesting only tree farm, or does it mean using whatever imputs are optimal to maintain continually high soil and crop productivity?

Mining is a good illustration of the slipperiness of the sustainability idea. No activity seems less sustainable than the extraction of minerals from the earth. Yet despite centries of mining, metals are as available as ever. The reason is that continual improvements in technology permit extraction of lower-grade deposits and increased recycling of metals.

The metals market is sustainable, even though high-grade ores have been used up, never to be replenished.

Of course there are impacts of mining other than resource exhaustion. Mining scars the face of the earth and pollutes the local environment, although these impacts can be minimized through regulation. Extraction and processing spread metals around the globe, raising concern about threats to health from toxic elements, such as mercury and lead. It is not clear whether mining these metals is ultimately sustainable. On the other hand, no one worries about health threats from the spread of common industrial metals like iron and aluminum, which have low toxicity and are abundant in the Earth's crust. There are incentives to recycle these metals, in order to save energy and minimize pollution, but none to stop using them.

Energy use raises the most dramatic issues of sustainability. Burning up the Earth's deposits of oil, gas, and coal is clearly an unsustainable practice. The energy cannot be recovered (second law of thermodynamics). Yet, is there something intrinsically bad about using up the fossil fuels? After all, they provide us with cheap, abundant energy, and useful commodities (e.g., plastics). The most important reason for curbing fossil fuel use is the rising level of heat-trapping CO_2, the end product of fossil fuel utilization. However, even this effect might be avoided by sequestering the CO_2 in the Earth or under the oceans.

In any event the use of fossil fuels as an energy source is ultimately unsustainable, but how long is ultimately? The time scale could be decades or centuries, depending on energy policies. To replace fossil fuels, we will have to turn to alternative energy sources: nuclear fission (if it can be developed safely), fusion (if it is techincally feasible), and/or renewable energy (the harnessing of which on large scale is difficult and expensive). Some people hope to end reliance on fossil fuels as soon as possible, by betting on renewable energy source now. But it is far from obvious how best to promote wise energy use while protecting the environment.

In sum, the sustainability concept is more a way of raising important questions than a guide to action. The answers involve making moral and political choices, based on a clear understanding of how the material world works. Things are rarely as simple as they seem.

16.2 Natural Energy Flows

It is instructive to view human energy use against a backdrop of the continual and massive flow of energy that occurs at the surface of the Earth. This flow is diagrammed in Figure 16.1; the magnitude of energy fluxes are given in units of 10^{20} kilojoules (kJ) per year. A tiny part of Earth's energy budget derives from non-solar sources: tidal energy (0.0013×10^{20} kJ), which arises from the gravitational attraction between the Moon and

Earth, and geothermal heat (0.01×10^{20} kJ), which emanates from Earth's molten core. (Human-generated nuclear energy is non-solar, but it is also minimal compared to total energy flows.) The rest of the energy on Earth comes from the sun, either directly or indirectly.

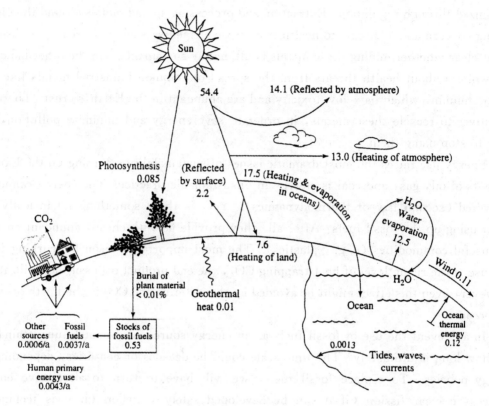

Figure 16.1 Annual energy fluxes on Earth (in 10^{20} kilojoules)

The sun radiates a nearly unimaginable number of kilojoules every year—1.17×10^{31}. A very small fraction of the total, 54.4×10^{20} kJ per year, is intercepted by Earth, which is 150 million kilometers (93 million miles) from the sun. Of this quantity, about 30 percent is reflected or scattered back into space from its atmosphere (26 percent) or surface (4 percent). This fraction is called the albedo, and it contributes significantly to the overall energy balance of Earth.

The rest of the light is absorbed by Earth's atmosphere (24 percent), land surface (14 percent), and oceans (32 percent), and is converted to heat before being re-radiated back into space. This heat flow drives Earth's weather system via wind, rain, and snow. About half of the absorbed energy reaching Earth's surface flows through the hydrological cycle, the massive evaporation and precipitation of water upon which we depend for our freshwater supplies. While it takes 4.2 joules (1 calorie) to heat a gram of water by 1℃, much more energy is needed to vaporize the same gram of water; the energy required to vaporize a liquid is called its latent heat of vaporization. At 15℃, which is the average annual global temperature, the latent heat of water is 2.46 kJ/g. The latent heat is released again when water vapor condenses into rain. This is the reason that rainfall is associated with storms; even a modest rainfall releases a huge amount of energy.

We extract a very small fraction of the energy in the hydrological cycle through dams and hydroelectric power generation, which constitute an indirect tapping of solar energy flow.

About 0.34 percent of the sunlight absorbed in Earth's surface (land plus oceans) is used by green plants and algae in photosynthesis. We depend on this fraction of the solar flux for our food supply and for a habitable earth. Some of the energy we use is provided by burning wood and other biomass forms (garbage, cow dung), and most of the rest is obtained by mining the store of photosynthetic products buried in ages past, in the form of fossil fuels.

16.3 Crisis of Atmospheric Chemistry

Not so long ago you would have been most unlikely to find in a book about chemistry any mention of the way in which it is relevant to our environment. Today, however, atmospheric and environmental chemistry are no longer regarded as arcane backwaters of science but as matters for immediate and global concern. Atmospheric scientists have rather suddenly found themselves at the focus of public and media attention, and their research determines the polices of governments. For the world has at least awoken to a fact that these scientists have always recognized: the chemical compostion of the atmosphere exerts a profound influence on our environment, and upsetting its delicate balance can have grave consequences for the planet.

Foremost amongst the problems to which atmospheric scientists must now address their attention are the threat of global warming (the so-called greenhouse effect), the deterioration of the ozone layer, the deleterious effects of acid rain and rising levels of pullutants such as lead, mercury and radioactive substances throughout the world. These issues have generated vigorous and sometimes bitter controversy, all the more so because their implications are of so much more than purely scientific interest. Industrial companies are having to face up to the potential damage that can be done to the environment by the chemicals that account for a part of their profits, while the increasing demand for power generation must now be weighted against the harmful effects of the waste gases or hazardous substances that are generally created in the process.

Apparently, atmospheric chemistry lies at the heart of some of these matters. A discussion of this kind cannot really be comprehended in its true context without some explanation, as far as we can yet provide it, of how our atmosphere got to be the way it is today. As we all know, the Earth has not always been blessed with a life-sustaining blanket. No, our present atmosphere did not beget life: life begat the present atmosphere. The atmospheric scientist James Lovelock has suggested that it is time to stop thinking of life, the solid Earth, the oceans and the atmosphere as independent systems, and to regard them instead as interconnected aspects of our planet. This point of view, which is central to Loverlock's "Caia" hypothesis, should help us to appreciate why we cannot divorce our activities from their environmental consequences, or assume that the environment has an infinite capacity to absorb our mess. Our atmosphere is a privilege that should not be taken for granted.

16.4 The Chemical Control of Climate: the Global Warmers

Of all the greenhouse gases, carbon dioxide is the most significant. Human activities have increased the amount of atmospheric CO_2 by about 26 percent since the Industrial Revolution, mainly through the burning of fossil fuels (coal, oil and natural gas). The destruction of vast areas of forest, particularly in South America, also has an important impact on atmospheric CO_2; forest act as natural "sponges" which soak up this gas from the atmosphere and fix the carbon in plant matter. If the trees are chopped down and burnt out to decompose, the carbon is released again as CO_2. Natural fluctuations of the carbon cycle can bring about substantial changes in concentrations of atmospheric CO_2, as the Vostok ice-core record indicates; the important question is how the natural sources and sinks (removal pathways) of CO_2 will change in the face of manmade (anthropogenic) inputs.

Although there is roughly two hundred times less methane than carbon dioxide in the atmosphere today, its potential role in global warming is significant. Molecule for molecule, CH_4 has a greater greenhouse warming effect than CO_2, as it is a stronger absorber of infrared radiation. Since the Industrial Revolution, the amount of methane in the atmosphere has doubled. A variety of human activities have contributed to this rise, largely connected with agriculture and land management. The cultivation of rice in paddy fields makes the biggest contribution: rise plants produce methane as they grow. Rice production, nearly all of which is practised in Asia, has approximately doubled since 1940. Ruminant animals such as cattle and sheep brew up considerable quantities of methane in their digestive systems, expulsion of which represents the second most important anthropogenic source. Burning of vegetation in tropical forests and savannahs releases methane, while other sources include the fermentation and decay of organic wastes in rubbish tips and landfills and the leakage of natural gas during coal mining and gas drilling and transmission along pipelines.

But there are substantial natural sources of methane too. Microbial processes in wetlands such as bogs, swamps and tundra produce roughly as much methane as paddy fields now do, while termite communities are thought to emit as much as is produced by burning of vegetation. Bilogical processes in oceans, lakes and rivers also produce small quantities of methane.

The most important sink for atmospheric methane is its chemical destruction in the atmosphere. The troposphere—that part of the atmosphere extending from ground level to an altitude of about 10~15 kilometers—contains significant quantities of reactive hydroxyl (OH) species. These attack methane to form a variety of products including carbon monoxide and water (both of which, however, themselves contribute to global warmings).

Chlorofluorocarbons are primarily a source of environmental concern because of their ozone-destroying properties (discussed later). But they are also very strong absorbers of infrared radiation, and so make a small but significant contribution to global warming despite being present in the atmosphere in far smaller concentrations than carbon dioxide or methane. The percentage of CFCs in the atmosphere is entirely mankind's doing—there are no natural sources. These gases are manufactured industrially for use as aerosol propellants, refrigerants, solvents and foam-blowing agents, uses that rely on their extreme inertness towards chemical reactions. But this unreactivity means that there are no effective removal pathways of CFCs in the atmosphere until they find their way up to the atmosphere and destroy ozone. The good news, however, is that CFCs are now so clearly a bad thing that the pressure on industry to replace them with less harmful compounds is considerable, and has resulted in agreements to phase out their use in the coming decades. It can be expected that the importance of CFCs as greenhouse gases will therefore decrease in the future. Ironically, the ozone-destroying capacity of CFCs diminishes their net global-warming effect, because ozone too is a greenhouse gas.

Nitrous oxide is the product of a wide range of biological processes in both the oceans and soils, although the details of these processes and the precise magnitudes of the natural sources and sinks are not well understood. Human activities have boosted the atmospheric concentration of nitrous oxide by about 8 per cent since industrial times, primarily by the burning of fossil fuels and forests and by the use of nitrogen-rich fertilizers (nitrates and ammonium salts).

Amongst the minor greenhouse gases are other nitrogen oxide, ozone and carbon monoxide. In contrast to its beneficial, UV-filtering role in the stratosphere (10~50 kilometers above the Earth's surface), at ground level ozone is not at all friendly to the environment—it is a hazardous, poisonous pollutant, harmful to the eyes and lungs and damaging to plants. Levels of troposphere ozone appear to have increased by as much as two or threefold relative to the past century as a result of fossil-fuel burning and industrial activity.

16.5 Green Chemistry and Development of Green Chemicals and Catalysts

The chemical industry has made many positive contributions to modern life, but these improvements have come at a price. The chemical products themselves, the chemicals used to produce them, and the byproducts of their production have sometimes been harmful to our health and the environment. Concern about these dangers has led to a movement in the chemical industry that is sometimes called green chemistry or environmentally benign chemistry. Its goal is to produce, process, and use chemicals in new ways that pose fewer risks to humans and their environment.

One such chemical named 4, 5-dichloro-2-n-octyl-4-isothiazolin-3-one, otherwise known as Sea-Nine® antifoulant, has been developed by the Rohm and Hass Company, which received the Green Chemistry Challenge's 1996 Designing Safer Chemicals Award. As soon as a new ship goes into the water, various plants and animals begin to grow on its bottom. This growth slows the ship's progress and causes other problems as well, ultimately costing the shipping industry an estimated $3 billion per year. Substances called antifoulants are painted on the bottoms of ships to prevent the unwanted growth, but most of them can be harmful to the marine environment. Sea-Nine® is a more environmentally friendly chemical than the alternatives because it breaks down in the ocean more quickly and does not accumulate in marine organisms.

Because catalysts are used by the chemical industry to produce many essential products, the discovery of new and better catalysts is important for meeting the goals of Green Chemistry. As catalysts improve, less energy will be required for the production of industrial chemicals, and raw material and byproducts will become more benign.

The Monsanto Company, as an example, received the 1996 Alternate Synthetic Pathways Award (a Presidential Green Chemistry Challenge Award) for improving the methods used in the production of their herbicide Rounduo®. Formerly, the synthesis of this product required the use of ammonia, formaldehyde, hydrogen cyanide, and hydrochloric acid. All of these substances pose dangers to the workers using them and to the environment, but hydrogen cyanide in particular is extremely toxic. Another problem was that overall process produced up to 1 kilogram of cyanide- and formaldehyde-containing waste for every 7 kilograms of product. When Monsato developed a process that used a new metallic copper catalyst for making Rounduo®, the company was able to eliminate cyanide and formaldehyde from the production and generate no waste at all.

Biocatalysts—catalysts produced by living organisms—are another important category of new catalysts being developed for industry. Enzymes are biocatalysts capable of forming very specific products very quickly. In some cases, their use in industrial reactions eliminates the need for hazardous reactions or the production of hazardous byproducts. For example, the conventional synthesis of acrylamide (used to make dyes, adhensives, and permanent-press fabrics) employs sulfuric acid in the first step of a two-step process and ammonia in the second step. Both of these substances can be difficult or even dangerous to handle. The conventional synthesis forms acrylamide and the waste product ammonium sulfate. An alternative technique, using an enzyme called nitrile hydratase, eliminates the need for sulfuric acid and ammonia and does not produce ammonium sulfate.

Old process
$$CH_2=CHCN + H_2O \xrightarrow{1.\ H_2SO_4;\ 2.\ NH_3} CH_2=CHCONH_2 + (NH_4)_2SO_4$$

New process
$$CH_2=CHCN + H_2O \xrightarrow{\text{Nitrile hydratase}} CH_2=CHCONH_2$$

第17章 药物化学术语
Chapter 17 Medicinal Chemical Terms

17.1 Scope of Medicinal Chemistry

Medicinal chemistry and pharmaceutical chemistry are disciplines at the intersection of chemistry, especially synthetic organic chemistry, and pharmacology and various other biological specialties, where it is involved with design, chemical synthesis and development for market of pharmaceutical agents (drugs).

Compounds used as medicines are most often organic compounds, which are often divided into the broad classes of small organic molecules (e. g. , atorvastatin, fluticasone, clopidogrel) and "biologics" (infliximab, erythropoietin, insulin glargine), the latter of which are most often medicinal preparations of proteins (natural and recombinant antibodies, hormones, etc.). Inorganic and organometallic compounds are also useful as drugs (e. g. , lithium and platinum-based agents such as lithium carbonate and cis-platin).

In particular, medicinal chemistry in its most common guise—focusing on small organic molecules—encompasses synthetic organic chemistry and aspects of natural products and computational chemistry in close combination with chemical biology, enzymology and structural biology, together aiming at the discovery and development of new therapeutic agents. Practically speaking, it involves chemical aspects of identification, and then systematic, thorough synthetic alteration of new chemical entities to make them suitable for therapeutic use. It includes synthetic and computational aspects of the study of existing drugs and agents in development in relation to their bioactivities (biological activities and properties), i. e. , understanding their structure-activity relationships (SAR). Pharmaceutical chemistry is focused on quality aspects of medicines and aims to assure fitness for purpose of medicinal products.

At the biological interface, medicinal chemistry combines to form a set of highly interdisciplinary sciences, setting its organic, physical, and computational emphases alongside biological areas such as biochemistry, molecular biology, pharmacognosy and pharmacology, toxicology and veterinary and human medicine; these, with project management, statistics, and pharmaceutical business practices, systematically oversee altering identified chemical agents such that after pharmaceutical formulation, they are safe and efficacious, and therefore suitable for use in treatment of disease.

17.2 Basic Medicinal Chemical Terms from IUPAC Recommendations(1998)

Term	Explanation
active transport	~ is the carriage of a solute across a biological membrane from low to high concentration that requires the expenditure of (metabolic) energy.
address-message concept	~ refers to compounds in which part of the molecule is required for binding (address) and part for the biological action (message).
ADME	~ is the abbreviation for Absorption, Distribution, Metabolism, Excretion.
affinity	~ is the tendency of a molecule to associate with another. The affinity of a drug is its ability to bind to its biological target (receptor, enzyme, transport system, etc.). For pharmacological receptors it can be thought of as the frequency with which the drug, when brought into the proximity of a receptor by diffusion, will reside at a position of minimum free energy within the force field of that receptor. For an agonist (or for an antagonist) the numerical representation of affinity is the reciprocal of the equilibrium dissociation constant of the ligand-receptor complex denoted K_A, calculated as the rate constant for offset (k_{-1}) divided by the rate constant for onset (k_1).
agonist	An ~ is an endogenous substance or a drug that can interact with a receptor and initiate a physiological or a pharmacological response characteristic of that receptor (contraction, relaxation, secretion, enzyme activation, etc.).
allosteric binding sites	~ are contained in many enzymes and receptors. As a consequence of the binding to allosteric binding sites, the interaction with the normal ligand may be either enhanced or reduced.
allosteric enzyme	An ~ is an enzyme that contains a region to which small, regulatory molecules ("effectors") may bind in addition to and separate from the substrate binding site and thereby affect the catalytic activity. On binding the effector, the catalytic activity of the enzyme towards the substrate may be enhanced, in which case the effector is an activator, or reduced, in which case it is a de-activator or inhibitor.
allosteric regulation	~ is the regulation of the activity of allosteric enzymes.
analog	An ~ is a drug whose structure is related to that of another drug but whose chemical and biological properties may be quite different.
antagonist	An ~ is a drug or a compound that opposes the physiological effects of another. At the receptor level, it is a chemical entity that opposes the receptor-associated responses normally induced by another bioactive agent.

Term	Explanation
antimetabolite	An ~ is a structural analog of an intermediate (substrate or coenzyme) in a physiologically occurring metabolic pathway that acts by replacing the natural substrate thus blocking or diverting the biosynthesis of physiologically important substances.
antisense molecule	An ~ is an oligonucleotide or analog thereof that is complementary to a segment of RNA (ribonucleic acid) or DNA (deoxyribonucleic acid) and that binds to it and inhibits its normal function.
autacoid	An ~ is a biological substance secreted by various cells whose physiological activity is restricted to the vicinity of its release; it is often referred to as local hormone.
autoreceptor	An ~, present at a nerve ending, is a receptor that regulates, via positive or negative feedback processes, the synthesis and/or release of its own physiological ligand.
bioassay	A ~ is a procedure for determining the concentration, purity, and/or biological activity of a substance (e.g., vitamin, hormone, plant growth factor, antibiotic, enzyme) by measuring its effect on an organism, tissue, cell, enzyme or receptor preparation compared to a standard preparation.
bioisostere	A ~ is a compound resulting from the exchange of an atom or of a group of atoms with another, broadly similar, atom or group of atoms. The objective of a bioisosteric replacement is to create a new compound with similar biological properties to the parent compound. The bioisosteric replacement may be physicochemically or topologically based.
bioprecursor prodrug	A ~ is a prodrug that does not imply the linkage to a carrier group, but results from a molecular modification of the active principle itself. This modification generates a new compound, able to be transformed metabolically or chemically, the resulting compound being the active principle.
biotransformation	~ is the chemical conversion of substances by living organisms or enzyme preparations.
CADD	~ is the the abbreviation for computer-assisted drug design.
carrier-linked prodrug (carrier prodrug)	A ~ is a prodrug that contains a temporary linkage of a given active substance with a transient carrier group that produces improved physicochemical or pharmacokinetic properties and that can be easily removed in vivo, usually by a hydrolytic cleavage.

第 17 章 药物化学术语(Medicinal Chemical Terms)

续表

Term	Explanation
cascade prodrug	A ~ is a prodrug for which the cleavage of the carrier group becomes effective only after unmasking an activating group.
catabolism	~ consists of reactions involving endogenous organic substrates to provide chemically available energy (e. g., ATP) and/or to generate metabolic intermediates used in subsequent anabolic reactions.
catabolite	A ~ is a naturally occurring metabolite.
clone	A ~ is a population of genetically identical cells produced from a common ancestor. Sometimes, "clone" is also used for a number of recombinant DNA (deoxyribonucleic acid) molecules all carrying the same inserted sequence.
codon	A ~ is the sequence of three consecutive nucleotides that occurs in mRNA which directs the incorporation of a specific amino acid into a protein or represents the starting or termination signals of protein synthesis.
coenzyme	A ~ is a dissociable, low-molecular weight, non-proteinaceous organic compound (often nucleotide) participating in enzymatic reactions as acceptor or donor of chemical groups or electrons.
combinatorial synthesis	~ is a process to prepare large sets of organic compounds by combining sets of building blocks.
combinatorial library	A ~ is a set of compounds prepared by combinatorial synthesis.
CoMFA	~ is the the abbreviation for Comparative Molecular Field Analysis.
comparative molecular field analysis (CoMFA)	~ is a 3D-QSAR method that uses statistical correlation techniques for the analysis of the quantitative relationship between the biological activity of a set of compounds with a specified alignment, and their three-dimensional electronic and steric properties. Other properties such as hydrophobicity and hydrogen bonding can also be incorporated into the analysis. (See also Three-dimensional Quantitative Structure-Activity Relationship [3D-QSAR].)
computational chemistry	~ is a discipline using mathematical methods for the calculation of molecular properties or for the simulation of molecular behaviour.
computer-assisted drug design (CADD)	~ involves all computer-assisted techniques used to discover, design and optimize biologically active compounds with a putative use as drugs.

Term	Explanation
congener	A ~ is a substance literally con-(with) generated or synthesized by essentially the same synthetic chemical reactions and the same procedures. Analogs are substances that are analogous in some respect to the prototype agent in chemical structure. Clearly congeners may be analogs or vice versa but not necessarily. The term congener, while most often a synonym for homologue, has become somewhat more diffuse in meaning so that the terms congener and analog are frequently used interchangeably in the literature.
cooperativity	~ is the interaction process by which binding of a ligand to one site on a macromolecule (enzyme, receptor, etc.) influences binding at a second site, e.g., between the substrate binding sites of an allosteric enzyme. Cooperative enzymes typically display a sigmoid (S-shaped) plot of the reaction rate against substrate concentration.
3D-QSAR	~ is the abbreviation for Three-dimensional Quantitative Structure-Activity Relationship.
de novo design	~ is the design of bioactive compounds by incremental construction of a ligand model within a model of the receptor or enzyme active site, the structure of which is known from X-ray or nuclear magnetic resonance (NMR) data.
distomer	A ~ is the enantiomer of a chiral compound that is the less potent for a particular action. This definition does not exclude the possibility of other effect or side effect of the distomer.
docking studies	~ are molecular modeling studies aiming at finding a proper fit between a ligand and its binding site.
double-blind study	A ~ is a clinical study of potential and marketed drugs, where neither the investigators nor the subjects know which subjects will be treated with the active principle and which ones will receive a placebo.
double prodrug (or pro-prodrug)	A ~ is a biologically inactive molecule which is transformed in vivo in two steps (enzymatically and/or chemically) to the active species.
drug	A ~ is any substance presented for treating, curing or preventing disease in human beings or in animals. A drug may also be used for making a medical diagnosis or for restoring, correcting, or modifying physiological functions (e.g., the contraceptive pill).

Term	Explanation
drug disposition	~ refers to all processes involved in the absorption, distribution, metabolism and excretion of drugs in a living organism.
drug latentiation	~ is the chemical modification of a biologically active compound to form a new compound, which in vivo will liberate the parent compound. Drug latentiation is synonymous with prodrug design.
drug targeting	~ is a strategy aiming at the delivery of a compound to a particular tissue of the body.
dual action drug	A ~ is a compound which combines two desired different pharmacological actions at a similarly efficacious dose.
efficacy	~ describes the relative intensity with which agonists vary in the response they produce even when they occupy the same number of receptors and with the same affinity. Efficacy is not synonymous to intrinsic activity. Efficacy is the property that enables drugs to produce responses. It is convenient to differentiate the properties of drugs into two groups, those which cause them to associate with the receptors (affinity) and those that produce stimulus (efficacy). This term is often used to characterize the level of maximal responses induced by agonists. In fact, not all agonists of a receptor are capable of inducing identical levels of maximal responses. Maximal response depends on the efficiency of receptor coupling, i.e., from the cascade of events, which, from the binding of the drug to the receptor, leads to the observed biological effect.
elimination	~ is the process achieving the reduction of the concentration of a xenobiotic including its metabolism.
enzyme	An ~ is a macromolecule, usually a protein, that functions as a (bio-)catalyst by increasing the reaction rate. In general, an enzyme catalyzes only one reaction type (reaction selectivity) and operates on only one type of substrate (substrate selectivity). Substrate molecules are transformed at the same site (regioselectivity) and only one or preferentially one of chiral a substrate or of a racemate is transformed (enantioselectivity (special form of stereoselectivity)).
enzyme induction	~ is the process whereby an (inducible) enzyme is synthesized in response to a specific inducer molecule. The inducer molecule (often a substrate that needs the catalytic activity of the inducible enzyme for its metabolism) combines with a repressor and thereby prevents the blocking of an operator by the repressor leading to the translation of the gene for the enzyme.

Term	Explanation
enzyme repression	~ is the mode by which the synthesis of an enzyme is prevented by repressor molecules. In many cases, the end product of a synthesis chain (e. g., an amino acid) acts as a feed-back corepressor by combining with an intracellular aporepressor protein, so that this complex is able to block the function of an operator. As a result, the whole operation is prevented from being transcribed into mRNA, and the expression of all enzymes necessary for the synthesis of the end product enzyme is abolished.
eudismic ratio	~ is the potency of the eutomer relative to that of the distomer.
eutomer	The ~ is the enantiomer of a chiral compound that is the more potent for a particular action. (See also distomer.)
genome	A ~ is the complete set of chromosomal and extrachromosomal genes of an organism, a cell, an organelle or a virus; the complete DNA (deoxyribonucleic acid) component of an organism.
hansch analysis	~ is the investigation of the quantitative relationship between the biological activity of a series of compounds and their physicochemical substituent or global parameters representing hydrophobic, electronic, steric and other effects using multiple regression correlation methodology.
hapten	A ~ is a low molecular weight molecule that contains an antigenic determinant but which is not itself antigenic unless combined with an antigenic carrier.
hard drug	A ~ is a nonmetabolizable compound, characterized either by high lipid solubility and accumulation in adipose tissues and organelles, or by high water solubility. In the lay press the term "Hard Drug" refers to a powerful drug of abuse such as cocaine or heroin.
heteroreceptor	A ~ is a receptor regulating the synthesis and/or the release of mediators other than its own ligand. (See also autoreceptor.)
homologue	The term ~ is used to describe a compound belonging to a series of compounds differing from each other by a repeating unit, such as a methylene group, a peptide residue, etc.
hormone	A ~ is a substance produced by endocrine glands, released in very low concentration into the bloodstream, and which exerts regulatory effects on specific organs or tissues distant from the site of secretion.
hydrophilicity	~ is the tendency of a molecule to be solvated by water.

Term	Explanation
hydrophobicity	~ is the association of non-polar groups or molecules in an aqueous environment which arises from the tendency of water to exclude non polar molecules. (See also lipophilicity.)
IND	~ is the abbreviation for Investigational New Drug.
intrinsic activity	~ is the maximal stimulatory response induced by a compound in relation to that of a given reference compound. (See also partial agonist.) This term has evolved with common usage. It was introduced by Ariëns as a proportionality factor between tissue response and receptor occupancy. The numerical value of intrinsic activity (alpha) could range from unity (for full agonists, i.e., agonist inducing the tissue maximal response) to zero (for antagonists), the fractional values within this range denoting partial agonists. Ariëns' original definition equates the molecular nature of alpha to maximal response only when response is a linear function of receptor occupancy. This function has been verified. Thus, intrinsic activity, which is a drug and tissue parameter, cannot be used as a characteristic drug parameter for classification of drugs or drug receptors. For this purpose, a proportionality factor derived by null methods, namely, relative efficacy, should be used. Finally, "intrinsic activity" should not be used instead of "intrinsic efficacy". A "partial agonist" should be termed "agonist with intermediate intrinsic efficacy" in a given tissue.
inverse agonist	An ~ is a drug which acts at the same receptor as that of an agonist, yet produces an opposite effect. Also called negative antagonists.
isosteres	~ are molecules or ions of similar size containing the same number of atoms and valence electrons, e.g., O^{2-}, F^-, Ne. (See also bioisostere.)
latentiated drug	(See drug latentiation.)
lead discovery	~ is the process of identifying active new chemical entities, which by subsequent modification may be transformed into a clinically useful drug.
lead generation	~ is the term applied to strategies developed to identify compounds which possess a desired but non-optimized biological activity.
lead optimization	~ is the synthetic modification of a biologically active compound, to fulfill all stereoelectronic, physicochemical, pharmacokinetic and toxicologic required for clinical usefulness.

Term	Explanation
lipophilicity	~ represents the affinity of a molecule or a moiety for a lipophilic environment. It is commonly measured by its distribution behaviour in a biphasic system, either liquid-liquid (e. g., partition coefficient in octan-ol/water) or solid/liquid (retention on reversed-phase high performance liquid chromatography (RP-HPLC) or thin-layer chromatography (TLC) system). (See also hydrophobicity.)
medicinal chemistry	~ is a chemistry-based discipline, also involving aspects of biological, medical and pharmaceutical sciences. It is concerned with the invention, discovery, design, identification and preparation of biologically active compounds, the study of their metabolism, the interpretation of their mode of action at the molecular level and the construction of structure-activity relationships.
metabolism	The term ~ comprises the entire physical and chemical processes involved in the maintenance and reproduction of life in which nutrients are broken down to generate energy and to give simpler molecules (catabolism) which by themselves may be used to form more complex molecules (anabolism). In case of heterotrophic organisms, the energy evolving from catabolic processes is made available for use by the organism. In medicinal chemistry the term metabolism refers to the biotransformation of xenobiotics and particularly drugs. (See also biotransformation; xenobiotic.)
metabolite	A ~ is any intermediate or product resulting from metabolism.
me-too drug	A ~ is a compound that is structurally very similar to already known drugs, with only minor pharmacological differences.
molecular graphics	~ is the visualization and manipulation of three-dimensional representations of molecules on a graphical display device.
molecular modeling	~ is a technique for the investigation of molecular structures and properties using computational chemistry and graphical visualization techniques in order to provide a plausible three-dimensional representation under a given set of circumstances.
mutagen	A ~ is an agent that causes a permanent heritable change (i. e., a mutation) into the DNA (deoxyribonucleic acid) of an organism.
mutual prodrug	A ~ is the association in a unique molecule of two, usually synergistic, drugs attached to each other, one drug being the carrier for the other and vice versa.

Term	Explanation
NCE	~ is the abbreviation for New Chemical Entity.
NDA	~ is the abbreviation for New Drug Application.
new chemical entity (NCE)	A ~ is a compound not previously described in the literature.
non-classical isostere	~ has the same meaning as bioisostere.
nucleic acid	A ~ is a macromolecule composed of linear sequences of nucleotides that perform several functions in living cells, e. g., the storage of genetic information and its transfer from one generation to the next DNA (deoxyribonucleic acid), the expression of this information in protein synthesis (mRNA, tRNA) and may act as functional components of subcellular units such as ribosomes (rRNA). RNA (ribonucleic acid) contains D-ribose, DNA contains 2-deoxy-D-ribose as the sugar component.
nucleoside	A ~ is a compound in which a purine or pyrimidine base is bound via a N-atom to C-1 replacing the hydroxy group of either 2-deoxy-D-ribose or of D-ribose, but without any phosphate groups. (See also nucleotide.) The common nucleosides in biological systems are adenosine, guanosine, cytidine, and uridine (which contain ribose) and deoxyadenosine, deoxyguanosine, deoxycytidine and thymidine (which contain deoxyribose).
nucleotide	A ~ is a nucleoside in which the primary hydroxy group of either 2-deoxy-D-ribose or D-ribose is esterified by orthophosphoric acid. (See also nucleoside.)
oligonucleotide	An ~ is an oligomer resulting from a linear sequence of nucleotides.
oncogene	An ~ is a normal cellular gene which, when inappropriately expressed or mutated, can transform eukaryotic cells into tumour cells.
orphan drug	An ~ is a drug for the treatment of a rare disease for which reasonable recovery of the sponsoring firm's research and development expenditure is not expected within a reasonable time. The term is also used to describe substances intended for such uses.
partial agonist	A ~ is an agonist which is unable to induce maximal activation of a receptor population, regardless of the amount of drug applied. (See also intrinsic activity.)

Term	Explanation
pattern recognition	A ~ is the identification of patterns in large data sets using appropriate mathematical methodologies.
peptidomimetic	A ~ is a compound containing non-peptidic structural elements that is capable of mimicking or antagonizing the biological action(s) of a natural parent peptide. A peptidomimetic does no longer have classical peptide characteristics such as enzymatically scissille peptidic bonds. (See also peptoids.)
peptoid	A ~ is a peptidomimetic that results from the oligomeric assembly of N-substituted glycines.
Pfeiffer's rule	~ states that in a series of chiral compounds the eudismic ratio increases with increasing potency of the eutomer.
pharmacokinetics	~ refers to the study of absorption, distribution, metabolism and excretion (ADME) of bioactive compounds in a higher organism. (See also drug disposition.)
pharmacophore (pharmacophoric pattern)	A ~ is the ensemble of steric and electronic features that is necessary to ensure the optimal supramolecular interactions with a specific biological target structure and to trigger (or to block) its biological response. A pharmacophore does not represent a real molecule or a real association of functional groups, but a purely abstract concept that accounts for the common molecular interaction capacities of a group of compounds towards their target structure. The pharmacophore can be considered as the largest common denominator shared by a set of active molecules. This definition discards a misuse often found in the medicinal chemistry literature which consists of naming as pharmacophores simple chemical functionalities such as guanidines, sulfonamides or dihydroimidazoles (formerly imidazolines), or typical structural skeletons such as flavones, phenothiazines, prostaglandins or steroids.
pharmacophoric descriptors	~ are used to define a pharmacophore, including H-bonding, hydrophobic and electrostatic interaction sites, defined by atoms, ring centers and virtual points.
placebo	A ~ is an inert substance or dosage form which is identical in appearance, flavor and odour to the active substance or dosage form. It is used as a negative control in a bioassay or in a clinical study.

Term	Explanation
potency	~ is the dose of drug required to produce a specific effect of given intensity as compared to a standard reference. Potency is a comparative rather than an absolute expression of drug activity. Drug potency depends on both affinity and efficacy. Thus, two agonists can be equipotent, but have different intrinsic efficacies with compensating differences in affinity.
prodrug	A ~ is any compound that undergoes biotransformation before exhibiting its pharmacological effects. Prodrugs can thus be viewed as drugs containing specialized non-toxic protective groups used in a transient manner to alter or to eliminate undesirable properties in the parent molecule. (See also double prodrug.)
QSAR	~ is the abbreviation for Quantitative Structure-Activity Relationships.
quantitative structure-activity relationships (QSAR)	~ are mathematical relationships linking chemical structure and pharmacological activity in a quantitative manner for a series of compounds. Methods which can be used in QSAR include various regression and pattern recognition techniques.
receptor	A ~ is a molecule or a polymeric structure in or on a cell that specifically recognizes and binds a compound acting as a molecular messenger (neurotransmitter, hormone, lymphokine, lectin, drug, etc.).
receptor mapping	~ is the technique used to describe the geometric and/or electronic features of a binding site when insufficient structural data for this receptor or enzyme are available. Generally the active site cavity is defined by comparing the superposition of active to that of inactive molecules.
second messenger	A ~ is an intracellular metabolite or ion increasing or decreasing as a response to the stimulation of receptors by agonists, considered as the "first messenger". This generic term usually does not prejudge the rank order of intracellular biochemical events.
site-specific delivery	~ is an approach to target a drug to a specific tissue, using prodrugs or antibody recognition systems.
soft drug	A ~ is a compound that is degraded in vivo to predictable non-toxic and inactive metabolites, after having achieved its therapeutic role.
SPC	~ is the abbrevitation for structure-property correlations.
structure-activity relationship (SAR)	~ is the relationship between chemical structure and pharmacological activity for a series of compounds.

续表

Term	Explanation
structure-based design	~ is a drug design strategy based on the 3D structure of the target obtained by X-ray or NMR.
structure-property correlations (SPC)	~ refers to all statistical mathematical methods used to correlate any structural property to any other property (intrinsic, chemical or biological), using statistical regression and pattern recognition techniques.
systemic	~ means relating to or affecting the whole body.
teratogen	A ~ is a substance that produces a malformation in a foetus.
three-dimensional quantitative structure-activity relationship (3D-QSAR)	A ~ is the analysis of the quantitative relationship between the biological activity of a set of compounds and their spatial properties using statistical methods.
topliss tree	A ~ is an operational scheme for analog design.
transition-state analog	A ~ is a compound that mimics the transition state of a substrate bound to an enzyme.
xenobiotic	A ~ is a compound foreign to an organism (xenos (Greek) = foreign).

17.3 Rehabilitation of Old Drugs and Development of New Ones

Imagine that you are a research chemist who has been hired by a large pharmaceutical company to develop a new drug for treating AIDS. How are you going to do it? Modern approaches to drug development fall into four general categories.

1. Old Drug, New Use

One approach is to do a computer search of all drugs to try to find one that can be put to a new use. For example, imagine you want to develop a drug for combating the lesions seen in Kaposi's sarcoma, an AIDS-related condition. These lesions are caused by the abnormal proliferation of small blood vessels. A list of all of the drugs that are thought to inhibit the growth of blood vessels might include some that are effective in treating Kaposi's sarcoma.

One of the drugs on that list is thalidomide, originally developed as a sedative by a German pharmaceutical company in the 1950s. It was considered a safe alternative to other sedatives, which are lethal in large doses, but when it was also used to reduce nausea associated with pregnant women's "morning sickness", it caused birth defects in the babies they were carrying. Thalidomide never did receive approval in the United States, and it was removed from the European market in the 1960s. About 10 000 children were born with imcompletely formed arms and legs as a result of thalidomide's effects.

Thalidomide is thought to inhibit the formation of limbs in the fetus by slowing the formation of blood vessels, but what can be disastrous for unborn children can be lifesaving for others. Today the drug is being used as a treatment for Kaposi's sacroma, and it may also be helpful in treating AIDS-related weight loss and brain cancer.

2. Old Drug, New Design

Another approach to drug development is to take a chemical already known to have a certain desirable effect and alter it slightly in hopes of enhancing its potency. The chemists at the Celgene Corporation have taken this approach with thalidomide. They have developed a number of new drugs that are similar in structure to thalidomide but appear to be 400 to 500 times more potent.

3. Rational Drug Design

In a third, more direct, approach often called rational drug design, the researcher first tries to determine what chemicals in the body are leading to the trouble. Often these chemicals are enzymes, large molecules that contain an active site in their structure where other molecules must fit to cause a change in the body. Once the offending enzyme is identified, isolated, and purified, it is "photographed" by X-ray crystallography, which reveals the enzyme's three-dimensional structure, including the shape of the active site. The next step is to design a molecule that will fit into the active site and deactivate the enzyme. If the enzyme is important for the replication of viruses like the AIDS virus or a flu virus, the reproduction of the virus will be slowed.

4. Combinatorial Chemistry

The process of making a single new chemical, isolating it, purifying in quantities large enough for testing is time-consuming and expensive. If the chemical fails to work, all you can do is start again and hope for success with the next. Thus chemists are always looking for ways to make and test more new chemicals more rapidly. A new approach to the production of chemicals, called combinatorial chemistry, holds great promise for doing just that.

Instead of making one new chemical at a time, the stratergy of combinatorial chemistry is to make and test thousands of similar chemicals at the same time. It therefore requires highly efficient techniques for isolating and identifying different compounds. One way of easily separating the various products from the solution in which they form is to run the reaction on the surface of tiny polymer beads that can be filtered from the reaction mixture after the reaction takes place. The beads must be tagged in some way so that the researcher can identify which ones contain which new substance. One of more novel ways of doing this is to cause the reaction to take place inside a tiny capsule from which a microchip sends out an identifying signal.

After a library of new chemicals has been produced, the thousands of compounds need to be tested to see which have desirable properties. Unfortunately, the procedures for testing large number of chemicals are often less than precise. One approach is to test all of them in rapid succession for characteristic that suggests a desired activity. A secondary library is then made with a range of structures similar to the structure of any substance that has that characteristic, and these new chemicals are also tested. In this way, the chemist can zero in on the chemicals that are most likely to have therapeutic properties. (Note again the connection between structure and properties.) The most likely candidates are then made in large quantities, purified more carefully, and tested in more traditional ways. Combinatorial chemistry has shown promise for producing pharmaceuticals of many types, including anticancer drugs and drugs to combat AIDS.

第三部分 (Part Ⅲ)

化学文献选讲
(Chemical Literature)

第 18 章 说明性短文
Chapter 18 Descriptive Short Articles

18.1 Laboratory Safety Rules

1. Follow Directions Exactly and Completely

The directions have been carefully designed to avoid all accidents. Deviation from these directions may cause injury to you or your neighbor.

2. When in Doubt Ask Your Instructor

Your instructor is in the laboratory to answer any questions that come up in the course of the quarter's work.

3. Do not Perform Unauthorized Experiments

One of the common sources of accidents is the performance of experiments that have not been explicitly assigned.

4. Use the Hood for Evaporation of Anything other than Water

Most other vapors are either corrosive, inflammable or toxic and should not be allowed to escape into the room.

5. Know the Location and the Use of Fire Extinguishers

6. Keep Desk Top Clean and Dry at All Times

It is impossible to distinguish by casual inspection water from corrosive chemicals.

7. Fire Polish All Glass Tubing

8. Glass Tubing vs. Stoppers

More accidents result from careless insertion of glass tubing through a stopper than from any other operation. To insert glass tubing into a stopper, ascertain first if the hole is large enough to accommodate the tubing. Thoroughly wet both tubing and stopper (conveniently accomplished by a stream of water). Hold stopper between thumb and forefinger, not in the palm of the hand. Grasp the tubing close to the end that is being inserted into the stopper, and twist in with slight rotary motion. Do not push or twist hard. It is best to wrap the tubing with several layers of a cloth towel. After the tubing emerges from the stopper it may be pulled rather than pushed.

9. Do not Look Down a Test Tube

Keep your face away from any chemicals being heated in open apparatus. Also be careful not to point any such open apparatus toward anyone else.

10. Sniff!! Do not Inhale!!

When attempting to detect chemical odors sniff gently. Most chemical vapors are at least irritating if not toxic.

11. Do not Pick up Hot Objects

Be sure that all apparatus is cool before picking it up in your fingers.

12. Always Pour Concentrated Acids into Water Carefully

Any concentrated reagent should be handled with care. Always pour concentrated sulfuric acid into water down the side of the container. Never pour water into concentrated sulfuric acid.

13. Wash it off with Water

If any chemical is splashed on you, wash it off with plenty of water. Then report to the instructor if further treatment is needed.

14. First Aid

Report accidents to the instructor, he can give first aid. If an instructor is not immediately available, stockroom personnel or the supervisory assistant can give first aid or, if necessary, take you to the infirmary.

15. Protect Clothing

Acids and bases destroy fabrics. it is wise to protect clothing by wearing a laboratory apron. Acids or strong bases spilled on clothing may be neutralized by applying sodium bicarbonate (baking soda). One may also use ammonium hydroxide to neutralize acids.

16. Always Wear Goggles in the Laboratory

17. Emergency Showers

In the event of burning clothing or a drenching spill of a harmful chemical the injured person should use an emergency shower at once. Each person should know the location of emergency showers in the laboratory.

18. Emergency Exits

Familiarize yourself with the location of all exits from the laboratory, and formulate a plan of escape in case of an emergency. Some rooms have wall panels which are easily kicked out in case of need.

18.2 Keeping Records

In research it is important that good records be kept; otherwise an unnecessary repetition of work result. In an industrial research laboratory the investigator is expected to keep understandable notes written in ink in a bound notebook. Observations and numerical data are written in the notebook as the work is in progress. Loose scraps of paper are not approved as places to record results. Each day's work is usually signed by the worker and dated. Frequently the signature of the investigator is witnessed by a colleague. One reason for keeping a good record is to have the information available for use by the company. Another reason is to be able to prove in court (in the event of a patent lawsuit) that the work in question was done and to establish the date.

The notes of the student in this course are to be kept in a way much like that just mentioned. A bound notebook called a "chemistry notebook" by the bookstores should be used. Each exercise should be recorded as a unit in the notebook (records of two exercises should not be mixed together even though both are worked upon at the same time). Each exercise should start on an unused page in the notebook. Records should be kept in ink. If this is not possible at the time, entries in pencil will be accepted. At the start of each day's work the date should be written.

For each exercise the following entries should be made: (1) title; (2) object of the exercise or a statement of what is to be accomplished; (3) a brief statement of the procedure if the procedure is not in this manual; (4) a record of observations and data (observations will normally be recorded just after a statement of the procedure). When possible, data for an exercise should be presented in a neat table. All observations and data should be recorded in the notebook as the work is in progress. Scraps of paper should not be used; (5) results clearly indicated so that the instructor can find them; (6) calculations made from the data to obtain the results just given. (These should be presented in an orderly manner which is easy to read. Preliminary calculations may be made on scraps of paper, in order that the calculations shown in the notebook may be neat); (7) discussion and estimation of errors in observation and the associated uncertainty in the final result; (8) any other information called for in the exercise.

In the event that an error is made in recording an observation or a datum the erroneous words or numbers should not be erased. They should be crossed out and the correct information written nearby. A judge looks with suspicion upon an erasure in a notebook used as evidence in a patent case. An instructor looks with suspicion upon an erasure in a student's notebook. He wants to see what was erased.

18.3 Use of a Balance

A good workman should always be sure that an instrument which he uses is in good order, and he should use the device in such a way that his results are reliable. The student has an opportunity to practice this principle when using a laboratory balance, for this instrument is sensitive and it may easily be thrown out of adjustment.

Weighing procedures are much simplified by the use of an electric balance. This balance works in much the same way as the double pan instrument, except that the weights are manipulated by means of internally located levers controlled with simple external dials. These dials, as well as other parts of the balance, must be handled carefully to avoid dislodging the weights from their proper locations.

The following procedure should be carefully followed when performing a weighing with the Sartorius balance. Figure 18.1 may be consulted to locate the various parts of the instrument discussed below.

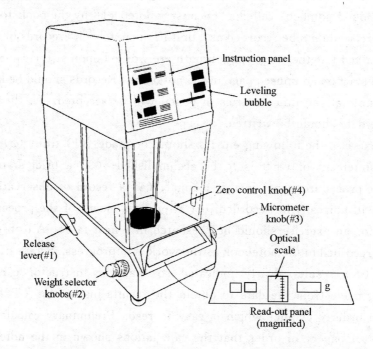

Figure 18.1 Sartorius Balance (Electric Balance)

1. Checking the Balance

(1) Check the leveling bubble. If the bubble is not within its circle, have the instructor level the balance.

(2) Be sure the pan is clean—there is a brush provided in the balance room for cleaning the pan.

(3) Set the mass dials (#2) and the micrometer dial (#3) to zero.

(4) Lower the pan control lever (#1) to release the beam.

(5) Use the zero adjustment (#4) to superimpose the zero line and the stationary reference line.

(6) Return lever (#1) to the neutral position.

2. Weighing a Sample

(1) Place the sample on the clean pan—be certain that the pan control is in neutral when the sample is placed on the pan.

(2) Raise level (#1) up to the coarse weighing position.

(3) The illuminated scale indicates the approximate weight of the object in grams. Set the mass dials (#2) to the correct weight in grams.

(4) Arrest the beam by returning lever (#1) to neutral.

(5) Release the beam by turning lever (#1) down slowly.

(6) Wait for the illuminated scale to come to a stop.

(7) Use the micrometer dial (#3) to move the scale so that one of the numbers on the illuminated scale coincides with the reference line.

3. Reading

(1) Always read from left to right.

(2) Note the position of the decimal point. Tens of grams and grams are shown to the left of the decimal; tenths and hundredths are read from the illuminated scale; thousandths and ten-thousandths appear in the micrometer dial window.

(3) The maximum weight is 100 g.

4. Completion

(1) Arrest the pan by turning lever (#1) to neutral—this turns the instrument off.

(2) Return the instrument dials to zero before leaving, and close the sliding windows on either side of the weighing chamber.

5. Cautions

(1) Do not place chemicals directly on the pan.

(2) Always handle the dials gently and slowly—otherwise, weights can be knocked off the balance beam.

6. Report to the Instructor

(1) Spillage inside the weighing chamber.

(2) Out-of-order instrument.

18.4 An Experiment

1. Preliminary

The chemistry laboratory is a place where you will learn by observation what the behavior of matter is. Forget preconceived notions about what is supposed to happen in a particular experiment. Follow directions carefully, and see what actually does happen. Be meticulous (very exact and careful) in recording the true observation even though you 'know' something else should happen. Ask yourself why the particular behavior was observed. Consult your instructor (teacher) if necessary. In this way, you will develop your ability for critical scientific observation.

2. Experiment I: Density of Solids

The density of a substance is defined as its mass per unit volume. The most obvious way to determine the density of a solid is to weigh a sample of the solid and then find out the volume that the sample occupies. In this experiment, you will be supplied with variously shaped pieces of metal. You are asked to determine the density of each specimen and then, by comparison with a table of known densities, to identify the metal in each specimen. As shown in Table 18.1, density is a characteristic property.

Table 18.1 Densities of Some Common Metals

Metals	Density/(g/cm^3)
Aluminium	2.7
Lead	11.4
Magnesium	1.8
Monel metal alloy	8.9
Steel (Fe, 1% C)	7.8
Tin	7.3
Wood's metal alloy	9.7
Zinc	7.1

3. Procedure

Procure (obtain) an unknown specimen from your instructor. Weigh the sample accurately on an analytical balance.

Determine the volume of your specimen by measuring the appropriate dimensions. For example, for a cylindrical sample, measure the diameter and length of the cylinder. Calculate the volume of the sample.

Determine the volume of your specimen directly by carefully sliding the specimen into a graduated cylinder containing a known volume of water. Make sure that no air bubbles are trapped. Note the total volume of the water and specimen.

Repeat with another unknown as directed by your instructor.

4. Questions

01 Which of the two methods of finding the volume of the solid is more precise? Explain.

02 Indicate how each of the following affects your calculated density: (a) part of the specimen sticks out of the water; (b) an air bubble is trapped under the specimen in the graduated cylinder; (c) alcohol (density, 0.79 g/cm^3) is inadvertently substituted for water (density, 1.00 g/cm^3) in the cylinder.

03 On the basis of the above experiment, devise a method for determining the density of a powdered solid.

04 Given a metal specimen from Table 1 in the shape of a right cone of altitude 3.5 cm with a base of diameter 2.5 cm. If its total weight is 41.82 g, what is the metal? ($V = \pi r^2 h/3$.)

18.5 Molecular Weight of a Substance

(1) Avogardro's Principle states that equal volumes of different gases at the same temperature and pressure contain the same number of molecules. The molecular weight of the vapor of a liquid (which boils below 100°C) will be determined by measuring its vapor density at the boiling point of water.

(2) Materials: gas density flask, rubber policeman, lead ring (from stockroom), unknown liquid (from instructor).

1. Procedure

(1) The gas density flask must first be dried, weighed, and filled with air.

Alternatively, the flask may be rinsed with a few mL of acetone and then dried by aspiration as described in paragraph (4). Do not heat the flask containing acetone with an open flame. Wipe the outside carefully with a clean towel, and then cap the flask with a clean, dry "rubber policeman". The policeman must be dry on the inside. This point must be checked with care. Weigh the capped flask as accurately as possible on the analytical balance and record the weight to four decimal places. The gas density flask is easily broken by mechanical shock, as in striking the floor or bench-top after being dropped, by falling or rolling over and hitting the glass tip on the bench-top, or by thermal shock resulting from water striking hot glass. It should be handled with care.

(2) Remove the cap and place on a clean watch glass. With a syringe, introduce into the density flask about 4 mL (*ca.* 1/2 the sample) of the unknown liquid. Place the flask in a 400 mL beaker, and hold it down with a lead ring or with a wire triangle bent so that it will wedge against the walls of the beaker and hold the bulb near, but not touching the bottom of the beaker. Fill the beaker with water to within an inch of the top. The open tip of the flask should extend about an inch or more above the water level. Heat the water to boiling. The liquid in the flask will have vaporized before the water reaches the boiling point. It is necessary to heat the remaining vapor so that its temperature comes to that of the boiling water. Heat the flask for at least five minutes after the onset of vigorous boiling to achieve this thermal equilibration. Quickly cap the flask with the "rubber policeman", and remove the burner. This must be done carefully to avoid splashing the boiling water or upsetting the beaker. Push down the cap (flask will press against the bottom of the beaker) until it slides about a quarter of an inch or more over the neck. The seal must be vacuum tight. The flask should not be lifted from the boiling water for inspection or for any other purpose until it has finally been capped by the policeman. Lifting the flask during boiling to see whether the liquid is gone is a serious experimental error.

(3) Remove the lead ring from the beaker (using crucible tongs and towel) and then remove the flask. Allow the flask to stand until it reaches room temperature (it may be cooled with tap water); then dry the outside carefully. Weigh the capped flask with its contents to four decimal places. Some liquid will be visible in the flask. This is formed by the condensation of vapor and is not an indication that liquid remained in (2). If time permits, the procedure in (2) and (3) should be repeated at this point to check the accuracy of the experiment. After performing each weighing the barometric pressure (see Appendix B) and the laboratory temperature should be measured.

(4) To calculate the vapor density, the volume of the flask must be determined. Fill the trough with cold water. With the tip under water, carefully remove the rubber cap from the density flask. As the cap is removed the partial vacuum created by the condensed vapor permits water to enter the flask. If the cap was placed on carefully, the flak will fill almost completely with water, and no more than a small bubble of air (due to vapor pressure of the unknown) will remain after the water stops entering the flask. Ask the instructor to fill the remaining vapor space by means of a syringe. After the water-filled vessel reaches room temperature, replace the rubber cap. Dry the flask and weigh it to the nearest tenth of a gram on a platform or triple beam balance (not the analytical balance). After completion of the experiment, water may be removed from the flask with the aspirator (allowing air to enter intermittently). Dry the flask by adding about 10 mL of acetone, shaking to distribute the acetone inside the flask and pumping (flask inverted) with the aspirator. Do not heat the flask. Leave the policeman off the flask when returning it to the stockroom.

(5) The analysis of the experimental data collected during the course of this experiment must include consideration of buoyant forces which act upon the flask when it is weighed. Any object which is weighed in air appears to have a smaller mass than it does when weighed in a vacuum. This mass deficit is attributable to the mass of air which the object displaces; there is a buoyant force acting upon the object which is directed upwards. (This buoyant force is what causes a lighter-than-air craft, such as a helium-filled balloon, to rise.)

Suppose that the mass of the evacuated glass flask together with the policeman has been determined by weighing the two in a vacuum, and let M_f be this mass. If the flask, containing a substances of mass M_s, is then weighed in air, the balance will give the apparent mass m_s as

$$m_s = M_f + M_s - V\rho_a$$

where V is the volume of air which the flask displaces and ρ_a is the density of air in the balance chamber. The term $V\rho_a$ is the effective mass of the buoyant force. If the density of air remains constant, i.e., temperature and pressure remain constant, from the weighing of one substance to the next, then differences between the various balance readings will be unaffected by buoyancy, or by the massive flask for that matter, and will be equal to the differences between the actual masses of the substances contained within the flask.

Additional information (density of air and water) needed to complete the calculations may be obtained. Using these densities and the weights of the flask filled with air and water, calculate the volume. With the latter known, the weight of the flask without air (weight in a vacuum) may be calculated and hence the weight and molecular weight of the unknown gas (at the boiling point of water).

2. Report

(1) The molecular weight of the unknown (calculations should be shown in the notebook).

(2) Molecular formula (instructor will furnish the percentage composition of the unknown) after the molecular weight has been reported.

(3) Theoretical molecular weight and percent error.

(4) Probable sources of experimental error.

(5) The calculation of the molecular weight assumed that the vapor in the flask behaved as an ideal gas. If the actual density of the unknown in the flask were 2% greater than that predicted by the gas law, what would the molecular weight be?

18.6 Tools of Chemistry: Nuclear Magnetic Resonance

A proton spins about its axis in the same way that an electron does, and the spinning produces a small magnetic moment. Because of this magnetic moment, the proton behaves like a tiny magnet: in an external magnetic field the proton aligns itself either parallel to the field or opposite, called antiparallel, to the field. Being parallel to the field is a somewhat lower energy condition than being antiparallel to the field. By allowing protons in a magnetic field to absorb energy in the radio frequency (rf) range of the electromagnetic spectrum, they can be made to change their alignment. This absorption is the basis for a spectroscopic technique.

Nuclear magnetic resonance (NMR) spectroscopy is the study of the structure of molecules as revealed by the absorption of radio frequency radiation by nuclei. When the nuclei being studied are protons, the technique is referred to as proton magnetic resonance (PMR). Some of the other nuclei that have net spin and can also be studied by nuclear magnetic resonance are deuterium (^2H), boron (^{11}B), carbon (^{13}C), and oxygen (^{17}O). Here we are interested only in the proton and what can be learned from its magnetic resonance.

The spectrum is measured by placing the sample, in solution, in an rf field of constant frequency and varying the strength of an applied magnetic field. At certain values of the applied magnetic field absorption of rf energy occurs, the alignment of the spin changes, and the energy absorption is detected and recorded.

The primary use of proton magnetic resonance spectroscopy is in the determination of the structure of organic compounds. The magnetic field strength at which a proton absorbs rf energy varies with the chemical environment of that proton.

Consider the methane and ethane molecules. The four protons in methane all have the same chemical environment, and so do the six protons in ethane. In propane, however, the protons have two different types of surroundings. Six protons have environment a (part of a $—CH_3$ group) and two protons have environment *b* (part of a —CH_2— group). In a nuclear magnetic resonance spectrum, methane and ethane each give only one peak, but propane gives two peaks. And, making NMR an even more useful technique, the areas under the two peaks in the propane spectrum are proportional to the number of protons of each type—they have the ratio 6 : 2.

The variation in applied magnetic field at which protons in different environments absorb rf energy is called the **chemical shift**. Since it is not possible to measure the absorption of rf energy by a free proton, chemical shift must be measured relative to a specific standard. Most often the standard is tetramethyl silane, TMS, a compound with only one type of H environment and a peak that appears conveniently near one end of the spectrum. The chemical shift of a particular peak is the difference between its absorption and that of TMS (the zero of the chemical shift scale). It is reported in parts per million, and called δ (delta):

$$\delta = \left(1 - \frac{B}{B_r}\right) \times 10^6$$

where B_r is the reference magnetic field strength (that at which absorption by TMS occurs) and B is the magnetic field strength at which absorption by the substance being studied is observed.

As an example of the kinds of information that can be obtained from an NMR spectrum, we examine the spectrum of 3-methyl-1-butene, given in left figure. Ignoring for a moment the splitting, or small peaks, in each group—the four areas of absorption show that there are four proton environments in the molecule, which is as it should be. The relative areas of 1 : 2 : 1 : 6 under the four groups of peaks show, as we can see in the structure, that there are six H's of one type (*a*), two of a second type (*d*), and one each of two other types (*b*, *c*). From the study of many spectra, the chemical shift ranges for different types of protons are known. Compare the following values with the spectrum: CH_3—C, 0.95~0.85; CH—, 1.6~1.4; HC=C, 4.8~6.2. With this information the complete structure of the compound could be worked out. The splitting into several peaks of the absorption peak for protons in each type of chemical environment is caused by the influence of the magnetic moment of the protons on adjacent atoms. With experience, additional information about structure can be obtained from the splitting.

The chemical shift values for hydrogen bonded to atoms other than carbon and in different types of chemical environments have been extensively tabulated. Since its discovery in 1946, NMR spectroscopy has rapidly become an important tool in organic structure determination and also for studying many types of inorganic compounds.

18.7 Tools of Chemistry: the Computer

It is widely believed that we are in the midst of an information revolution that will change society as profoundly as the industrial revolution did nineteenth century Europe and America. The digital computer is both the tool and the symbol of this revolution. Every scientific discipline has been affected by advances in computer technology, and chemistry is no exception.

The earliest applications of computational methods in chemistry were made by quantum theorists, who were trying to find accurate methods of calculating the electronic structure of molecules. Both the internal electronic energy of molecules and the distribution of electronic charge in a molecule are obtained by solving the Schrödinger equation. This equation can be solved exactly only for a few very simple problems of interest in chemistry, for example, the electronic structure of the hydrogen atom. It is from these exact solutions that our insight into chemical structure has developed; the concept of atomic orbitals comes directly from the solutions of the Schrödinger equation for the hydrogen atom. But even for the hydrogen molecule, an exact solution of the Schrödinger equation is impossible without the use of numerical methods. Chemists during the 1930s were using analog computers (and even mechanical desk calculators) to solve the Schrödinger equation numerically, and they were able to determine the bond energy of the hydrogen molecule to within 10 percent of the presently accepted experimental value.

For any molecule heavier than the hydrogen molecule, the Schrödinger equation is complex enough to render numerical computations very tedious if done by hand. As digital computers became available in universities during the 1950s, computational methods for solving Schrödinger's equation for molecules with many electrons were developed. There are now many research groups in both universities and industries that have computers dedicated entirely to electronic structural calculations, and accurate studies can be done on large, chemically interesting molecules. For example, it is now possible to compute the interaction energies between two molecules such as the biochemically important cyclic bases adenine and thymine when they are hydrogen-bonded, as in the double-helix form of DNA.

Of course, molecular structure can be determined by many experimental methods as well, for example, by X-ray diffraction, molecular spectroscopy, or magnetic resonance. In each of these experimental methods, the availability of computer-based methods of analyzing data has made possible the determination of structures for more complex than could have been done 20 years ago. For example, the complete three-dimensional structures of biologically significant molecules such as myoglobin, an oxygen-carrying protein that is similar to hemoglobin and contains more than one hundred atoms, has been completely worked out using X-ray diffraction and computer reduction of data.

Complex molecular structures can also be displayed on a variety of output devices, using the techniques of computer graphics. By representing the positions of all atoms within a molecule by vectors, each atom can be located in space with three numbers. Atoms themselves can be depicted in many ways: as points, as spheres with radii equal to the van der Waals radii of the atoms, or even as contour diagrams of electron density. Devices such as high resolution graphics terminals or plotters can then be used to draw a picture of the molecule, viewed from any angle at any distance. The viewing perspective can be changed until a particular structural feature of interest comes into view. The figures in the margin show examples of such computer-generated structures.

In addition, chemists have found that microcomputers can be used for more than driving beauties around video screens in arcade games: they can also be used very effectively in the laboratory to control experimental equipment and to help record data. For example, the pH electrode you read about in this chapter can be interfaced with a microcomputer to produce an instrument that will record the pH every second during a titration. With another electronic interface and some flow control valves, the microprocessor can be used to add specific volumes of titrant at regular intervals. With some additional programming, the two functions—data acquisition and experimental control—can be combined so that the volume of titrant added decreases as the endpoint is approached. It is then a simple matter to have the microcomputer calculate the molarity of the unknown acid. Now that a machine can do this much of your acid-base titration lab, what do you think the next step will be?

The real power of computer-based chemical instrumentation is seen in more complex experiments than titrations, however. For example, the tools of gas chromatography and mass spectrometry can be combined, with the aid of a minicomputer, to produce an analytical instrument of great sensitivity and specificity. A mixture of many different chemical compounds can be separated with chromatography, but simple detectors cannot identify the compounds as they are eluted from the column. A mass spectrometer can be used to read the "fingerprint" of each compound as it emerges. Instruments that combine mass spectrometers with chromatographs require delicate adjustments both before and during an experiment. Minicomputers are used to control the acquisition and analysis of the data, allowing specific compounds present in the original sample in concentrations of less than one part per million to be detected.

In the chemical industry, many chemical processes also require elaborate control of conditions such as temperature, pressure, flow rate, and catalyst condition, as well as immediate analyses of the process stream for the desired product. Computers are an essential tool used by chemical engineers in the design and operation of chemical plants.

The ability of computers to search through huge data bases for desired information is also making significant changes in the field of chemistry. From the graphical input of the structure of a molecule, the data base of existing compounds can be checked to see if a molecule of interest has been made. It is also possible to search the data base for other compounds containing structural features similar to those of the compound originally entered. With the knowledge gained in learning to deal with complex chemical structures in this way, another interesting use of computers has emerged. There are now programs that will aid organic chemists in designing successful syntheses of compounds that have not yet been made. Several of these programs use concepts developed in the field of artificial intelligence to guide the synthesis design.

//
第 19 章 期刊论文
Chapter 19　Periodical Papers

19.1　Progesterone from 3-Acetoxybisnor-5-cholenaldehyde and 3-Ketobisnor-4-cholenaldehyde[①]

The preparation of derivatives of bisnor-5-cholenaldehyde by the ozonization of the corresponding stigmasterol derivatives[01] has been extended to the ozonization of stigmastadienone[02], from which a yield of no less than 60% of 3-ketobisnor-4-cholenaldehyde (Ⅰ) was readily obtained. The amount of ozone used in this reaction was carefully controlled to obtain a maximum yield of aldehyde and a minimum yield of 3-ketobisnor-4-cholenic acid[03] along with some easily recoverable starting material.

This new aldehyde (Ⅰ) and 3β-acetoxybisnor-5-cholenaldehyde[01] (Ⅱ) have now been successfully degraded to progesterone by a method first used by Semmler[04] who showed that aldehydes, including eksantalol, phenylacetaldehyde, citronellal, citral and others, which possess one or two labile hydrogen atoms on the carbon adjacent to the aldehyde group, may be converted upon heating under reflux with acetic anhydride and sodium acetate into an unsaturated ester designated as an enol acetate. In the case of eksantalol he described a method of degrading the side chain; by ozonization of the enol acetate a new aldehyde or ketone which is one carbon poorer results.

We have found that when 3β-acetoxybisnor-5-cholenaldehyde (Ⅱ) was heated under reflux with acetic anhydride and sodium acetate for periods varying from six to twenty-four hours, the nicely crystalline unsaturated enol acetate (Ⅴ) could be isolated in yields of 70%. This enol acetate (Ⅴ), as such or as the C-5,6 dibromide, was readily ozonized to give 5-pregnene-3β-ol-20-one acetate (Ⅵ) which, upon saponification and oxidation, is readily convertible into progesterone (Ⅶ). Somewhat better over-all yields of pregnenolone acetate from the aldehyde were obtained when the enol acetate (Ⅴ) was not isolated and the double bond in the nucleus was left unprotected by bromine during the ozonolysis.

$$\text{Ⅰ} \xrightarrow{Ac_2O,\ NaAc} [\text{Ⅲ} + \text{Ⅳ}] \xrightarrow{O_3,\ then\ H^+} \text{Ⅶ}$$

$$\text{Ⅱ} \xrightarrow{Ac_2O,\ NaAc} \text{Ⅴ} \xrightarrow{O_3} \text{Ⅵ} \xrightarrow{1.\ Hydrolysis;\ 2.\ Oxidation} \text{Ⅶ}$$

---

① Heyl F W and Herr M E. *J. Am. Chem. Soc.* 72, 2617 (1950). The Research Laboratories of The Upjohn Co.

I : $R_1 =$ —CH(CH$_3$)—CHO
III : $R_1 =$ —C(CH$_3$)=CHOAc
VII : $R_1 =$ —COCH$_3$

II : $R_2 =$ —CH(CH$_3$)—CHO
IV : $R_2 =$ —C(CH$_3$)=CHOAc

V : $R_3 =$ —C(CH$_3$)=CHOAc
VI : $R_3 =$ —COCH$_3$

Structural Formulas for Compound I ~ VII

When 3-ketobisnor-4-cholenaldehyde (I) was heated under reflux with acetic anhydride and sodium acetate the pure intermediate enol acetate was not obtained. This product is very likely a mixture of di- and mono-enol acetates (III and IV). It was found, however, that this crude mixture could be very conveniently ozonized directly to progesterone (VII) in 60% over-all yield from the keto-aldehyde (I).

Experimental[5]

3-Ketobisnor-4-cholenaldehyde (I).—An ice-cooled solution of 4.11 g. (0.01 mol) of stigmastadienone[02], mp 124~125°C, in 250 mL of chloroform and 5 mL of pyridine was ozonized for thirty-eight minutes by passing through a stream of ozone-oxygen (25 mg of ozone per minute). The solvent was removed under nitrogen *in vacuo* and the colorless sirup residue taken up in 25 mL of glacial acetic acid and 50 mL of ether. Four grams of zinc dust was added while shaking over a period of ten minutes, the mixture diluted with 300 mL of ether and filtered. The filtrate was washed twice with water followed by cold 10% sodium hydroxide (at which point an insoluble sodium salt separated) and finally with water. The ether solution after drying over sodium sulfate was evaporated and the residue crystallized from isopropyl ether. The keto-aldehyde which separated in needles was collected on a Buchner funnel and washed with a little cold isopropyl ether; yield 1.95 g (60%), mp 148~151°C. A sample for analyses was crystallized from ether to constant mp 160~161°C.

Anal. Calcd. for $C_{22}H_{32}O_2$: C, 80.43; H: 9.82. Found: C, 80.69; H, 10.02; $[\alpha]^{24}D +82.5°$ (0.1252 g made up to 10 mL with chloroform, $[\alpha]^{24}D +1.032°$, l, 1 dm).

The residue from the isopropyl ether filtrate, upon crystallization from dilute acetone, gave 0.73 g of starting material, mp 121~123°C. If this recovered material is taken into account the yield of aldehyde is 72%.　　　　[N. B. Other paragraphs omitted]

References and Notes

[01]　Heyl, Centolella and Herr, This Journal, 69, 1957 (1947); 70, 2953 (1948)

[02]　Stigmastadienone was prepared by the Oppenauer oxidation of stigmasterol; Fernholz and Stavely, *ibid.*, 61, 2956 (1939)

[03]　This keto-acid has previously been prepared by chromic acid oxidation of 3- hydroxybisnor-5-cholenic acid dibromide; Butenandt and Mamoli, *Ber.*, 68B, 1857 (1935)

[04]　Semmler, *ibid.*, 42, 584, 962, 1161, 2014 (1909); Semmler and Schossberger, *ibid.*, 44, 991 (1911); Bedoukian, This Journal, 66, 1325 (1944); *ibid.*, 67, 1430 (1945); see also Bergmann and Stevens, *J. Org. Chem.*, 13, 10 (1948)

[05]　Melting points are corrected. Analyses and rotations were carried out by personnel of the Upjohn Microanalytical Laboratory

19.2　Exact Shapes of Random Walks in Two Dimensions[①]

Since the random walk problem was first presented by Pearson[01] in 1905, the shape of a walk which is either completely random or self-avoiding has attracted the attention of generations of researchers working in such diverse fields as chemistry, physics, biology and statistics[02~18]. Among many advances in the field made in the past decade is the formulation of the three-dimensional shape distribution function of a random walk as a triple Fourier integral plus its numerical evaluation and graphical illustration[11,18~20]. However, exact calculations of the averaged individual principal components of the shape tensor for a walk of a certain architectural type including an open walk have remained a challenge. Here we provide an exact analytical approach to the shapes of arbitrary random walks in two dimensions. Especially, we find that an end-looped random walk surprisingly has an even larger shape asymmetry than an open walk.

The shape of a random walk taking place in a d-dimensional space is often described by d principal components arranged in descending order, i.e., $S_1 \geqslant S_2 \geqslant \cdots \geqslant S_d$, of the shape tensor $\mathbf{S}$[07,11,15,17]. This tensor $\mathbf{S}$ is related to the inertia tensor $\mathbf{I}$ of an n-vertex walk by the equality $\mathbf{I} = n(s^2 \mathbf{1} - \mathbf{S})$ where unit mass is assumed for each vertex, $\mathbf{1}$ is the identity matrix, and s^2 is the trace of $\mathbf{S}$, i.e., $s^2 = \mathrm{tr}(\mathbf{S})$ with s historically termed the radius of gyration[03,05], i.e., the square root of the arithmetic mean of n squared distances of the vertices from their center of mass. For convenience, unit step length is further assumed for the walk.

[①]　Gaoyuan Wei; *Physica* A 222, 152~154 (1995). Department of Chemistry, Peking University, Beijing 100871, P. R. of China.

A walk may be either completely random or self-intersecting[01~03,05] (the large-n limit of the gaussian model) or self-avoiding (Edwards model)[02,04,10,12,15,16] and may have different architectural types[21] usually specified by the architecture or Kirchhoff matrix **K**. Let **Λ** denote the diagonal matrix of all $n-1$ nonzero eigenvalues of **K** times n^2. The eigenpolynomials of **Λ**, i.e., $P_{n-1}(x) = |\mathbf{1} + x\mathbf{\Lambda}^{-1}|$ for an end-looped or dumbbell-like walk, i.e., two identical large rings connected by a doubly-sized chain, may be written down with the use of graph theory, with the result that $D(x) = P_\infty(x^2) = U(x) U^2(x/2) B(x, -1/3) B(x, 2/3)$, where $U(x) = 4\operatorname{sh}(x/4)/x$ and $B(x,a) = [\operatorname{ch}(x/4) - a]/(1-a)$. From the above eigenpolynomial, one can obtain an analytic expression for the function $S_m(x)$ defined as the large-n limit of $\operatorname{tr}(\mathbf{\Lambda} + x^2 \mathbf{1})^{-m}$.

For arbitrary random walks in two dimensions, we find by using the method of Solč and Gobush[09] that shape factors, i.e., the δ_α defined as $<S_\alpha>/<s^2>$ which is the ratio of the averaged principal component of **S** to the mean square radius of gyration, and shape variance factors, the σ_α defined as $(<S_\alpha^2> - <S_\alpha>^2)/<s^2>^2$, are given by $\delta_\alpha = 1/2 + (-1)^\alpha \chi_1$ and $\sigma_\alpha = (1+3\mu_2)/4 + (-1)^\alpha \chi_2 - \delta_\alpha^2$, respectively. Here, $\mu_m = S_m(0)/S_1^m(0)$ and χ_m is defined as

$$\chi_m = S_1^{-m}(0) \int_0^\infty |x D(x+\mathrm{i}x)|^{-1} \operatorname{Im}[F_m(x+\mathrm{i}x)]\, \mathrm{d}x,$$

with $F_1(x) = S_1(x)$, $F_2(x) = S_2(x) + 2^{-1} S_1^2(x)$ and $\operatorname{Im}(x+\mathrm{i}y)$ denoting the imaginary part of $x+\mathrm{i}y$. Similarly, we can write down expressions for δ_α and σ_α for the $d=3$ case. However, a complication occurs in this case as it involves triple integrals over the restricted domains of the rotation group SO(3), which are difficult to evaluate accurately even by numerical means. Therefore, for random walks in a space with $3 \leqslant d < \infty$ and for self-avoiding walks, the exact evaluation of δ_α or σ_α remains a challenge. Numerical evaluations of δ_α and σ_α for two common types and one new type of random walks in two dimensions, i.e., open, closed and end-looped walks, based on the above general formulas have been made and the results are tabulated in Table 19.1. We note that our general formulas reproduce the earlier results for a closed random walk[09].

From Table 19.1, we find that the simulation results of Bishop and Michels[22] for shape factors of 2D chains and rings of finite length ($n=64$), i.e., 0.839 and 0.161 for chains and 0.755 and 0.245 for rings, are very close to our exact values for open and closed random walks; and that shape variance factors are in descending order, i.e., $\sigma_1 > \sigma_2$, for all three types of random walks, implying a broader distribution of the largest principal component. For an end-looped random walk, it is seen from Table 19.1 that it is more elongated than other types of random walks and even more asymmetrical than an open random walk though its average size is smaller than that of the latter (with a shrinking factor, i.e., the ratio of its mean square radius of gyration to that of the open walk,

Table 19.1 Shape and shape variance factors for open(1), closed(2) and end-looped(3) random walks in two dimensions

Type	δ_1	δ_2	σ_1	σ_2
1	0.832938	0.167062	0.369214	0.009090
2	0.754323	0.245677	0.155901	0.014739
3	0.852352	0.147648	0.439593	0.006567

of $51/64 \approx 0.796875$). This large shape asymmetry of the end-looped random walk may have important implications for the improvement of the rheological properties of end-looped linear polymers yet to be made or discovered.

Acknowledgements This work was supported in part by Peking University Excellent Young Scholars Fund and by National Natural Science Foundation and State Education Commission of China.

References

[01] Pearson K. *Nature* 72, 294~294; 342~342 (1905)
[02] Kuhn W. *Kolloid-Z.* 68, 2~15 (1934)
[03] Fixman M. *J. Chem. Phys.* 36, 306~310 (1962)
[04] Edwards S F. *Proc. Phys. Soc. (London)* 85, 613~624 (1965)
[05] Flory P J. Statistical Mechanics of Chain Molecules (Wiley-Interscience, New York, 1969)
[06] Solč, K & Stockmayer W H. *J. Chem. Phys.* 54, 2756~2757 (1971)
[07] Solč, K. *J. Chem. Phys.* 55, 335~344 (1971)
[08] Mazur J, Guttman C M & McCrackin F L. *Macromolecules* 6, 872~874 (1973)
[09] Solč, K & Gobush W. *Macromolecules* 7, 814~823 (1974)
[10] deGennes P G. Scaling Concepts in Polymer Physics (Cornell University Press, Ithaca, New York, 1979)
[11] Eichinger B E. *Macromolecules* 18, 211~216 (1985)
[12] Doi M & Edwards S F. The Theory of Polymer Dynamics (Oxford University Press, Oxford, 1986)
[13] Lawley D N. *Biometrika* 43, 128~136 (1956)
[14] Muirhead R J. Aspects of Multivariate Statistical Theory (Wiley, New York, 1982)
[15] Aronovitz J A & Nelson D R. *J. Physique* 47, 1445~1456 (1986)
[16] Freed K F. Renormalization Group Theory of Macromolecules (Wiley, New York, 1987)
[17] Rudnick J & Gaspari G. *Science* 237, 384~389 (1987)
[18] Wei G & Eichinger B E. *Macromolecules* 23, 4845~4855 (1990)
[19] Wei G & Eichinger B E. *Comput. Polym. Sci.* 1, 41~50 (1991)
[20] Wei G & Eichinger B E. *Ann. Inst. Statist. Math. (Tokyo)* 45, 467~475 (1993)
[21] Kuchanov S I, Korolev S V & Panyukov S V. *Adv. Chem. Phys.* 72, 115~326 (1988)
[22] Bishop M & Michels J P J. *J. Chem. Phys.* 85, 1074~1076 (1986)

19.3 Modelling of Molecular Networks with Negative Poisson's Ratios[①]

Naturally occurring materials have a possitive Poisson's ratio[01]: when stretched, the material becomes thinner. Recently, synthetic materials have been processed that exhibit a negative Poisson's ratio (becoming fatter when stretched)[02~06]. Such materials can have improved mechanical properties, such as enhanced shear moduli, indentation resistance and fracture toughness[02,07]. All these materials exhibt negative Poisson's ratio as a result of microstructures, or geometric units, that are at least tens of micrometres in size. We have used molecular modelling techniques to design molecular network with a negative Poisson's ratio. By altering the geometry of the repeat unit, Poisson's ratio can be varied to produce positive or negative values with mechanical properties that are either isotropic or anisotropic.

The Poisson's ratio of a material is given by $\nu_{xy} = \varepsilon_y/\varepsilon_x$, where ε_x is an applied tensile strain and ε_y is the resulting tensile strain in the transverse direction. For isotropic materials $\nu_{xy} = \nu_{yx} = \nu$ with the limits $-1 < \nu < 1/2$. For anisotropic materials neither of these limits apply. For example, orthotropic materials have Poisson's ratios anywhere within the range $|\nu_{xy}| < (E_x/E_y)^{1/2}$, where E_x and E_y are Young's moduli in the orthogonal direction x and y[08,09].

Examples of materials that have been fabricated with a negative Poisson's ration are two-dimensional honeycomb[04], three-dimensional foams[02,03] and microporous polymers[05,06,10]. An example of the simplest microstructure that produces this effect is illustrated schematically in Figure 19.1(a). A conventional hexagonal honeycomb has a positive Poisson's ratio, but making the cells "re-entrant", as in the figure, produces a negative Poisson's ratio.

The honeycombs and foams manufactured so far have low densities and are intrinsically weak and flexible. By reproducing this geometric unit on the molecular scale, however, it should be possible to take advantage of the innate free volume found, for example, in polymeric structures to increase the density and absolute stiffness of the material. Figure 19.1(b) illustrates a molecular network that reproduces, on the molecular scale, the features of the geometry in Figure 19.1(a). We suggest non-systematic naming convention for the hexagonal network of the form (n,m)-flexyne for the honeycomb structure, where n is the number of acetylene links on the diagonal branches and m is the number of links on the vertical branches, and (n,m)-reflexyne for the re-entrant structure. By altering the values of n and m, the structure can be either

---

① Evans K E, Nkansah M A, Hutchinson I J, Rogers S C. Nature 353, 124 (1991). Univesity of Liverpool and ICI Chemicals and Polymers Ltd., UK. (N. B. The original title of this scientific correspondence is Molecular Network Design.)

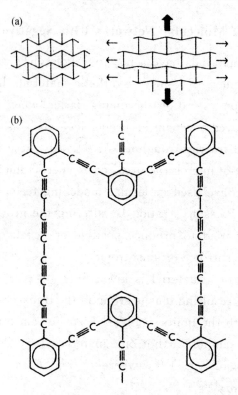

Figure 19.1 (a) Two-dimensional re-entrant honeycomb, showing transverse expansion on stretching; (b) (1,4)-reflexyne with a negative Poisson's ratio

isotropic or anisotropic with varying Poisson's ratios. The use of benzene rings at the junctions and acetylene groups along the branch creates, in principle, a planar two-dimensional structure. Although such a network may be hard to synthesize, it provides a simple example for molecular tailoring of Poisson's ratio.

A model has been developed[11] that successfully predicts Poisson's ratio for macroscopic honeycombs. This model assumes that deformation occurs by flexure, with no stretching of the cell arms, whereas in the molecular structure of Figure 19.1(b) deformation will occur by flexure, stretching and rotation of arms. We have used a molecular-mechanics program[12] which incorporates a standard valence force field to model the deformation of this structure and thus to calculate Poisson's ratio. Because of the symmetry of the network, only one uint cell need be modelled to determine the network's deformation behaviour. A force is applied in the x direction and the resulting displacements in the x and y directions are used to obtain ν_{xy}. Similarly, a force in the y direction is used to obtain ν_{yx}. Strains of less than 5% are used to ensure elastic behaviour.

In the table we give values for five (n,m)-reflexyne networks. The re-entrant structure does indeed produce negaive Poisson's ratios, although in general the simple flexure model over-estimates Poisson's ratio relative to the molecular-mechanics approach. This is because both stretching and rotational deformation will tend to increase the longitudinal deformation at the expense of the transverse. The value of Poisson's ratio can be altered by changing the shape of the repeat unit, $\nu_{xy}=1$ represents a truly isotropic structure. $\nu_{xy}=-1$ is only quasi-isotropic in that Poisson's ratios measured by stretching along directions away from the principal axes will have different values; it is a square-symmetric structure.

Table 19.2 Calculatel Values of Poisson's Ratios for (n,m)-Reflexyne

		Flexure model			Molecular mechanics		
n	m	ν_{xy}	ν_{yx}	ν_{xy}/ν_{yx}	ν_{xy}	ν_{yx}	ν_{xy}/ν_{yx}
2	6	−0.96	−1.04	0.93	−0.84	−0.81	1.04
2	8	−0.95	−1.05	0.92	−0.62	−1.14	0.55
1	4	−0.93	−1.08	0.86	−0.59	−0.83	0.72
1	6	−0.63	−1.58	0.40	−0.43	−1.14	0.38
1	8	−0.51	−1.96	0.26	−0.31	−1.47	0.21

The molecular network described here represents a first attempt at designing a material that demonstrates a negative Poisson's ratio owing to mechanisms acting at the molecular level. We are now investigating more complex structures that are more amenable to synthesis that have been proposed by Wei G Y and Edwards S F of Cambridge University. To avoid the cumbersome phrase "negative-Poisson-ratio materials", we suggest that they be called auxetic materials, or auxetics (from the Greek auxetos: that may be increased, referring to the width and volume increase when stretched).

References and Notes

[01] Landau L D & Lifshitz E M. Theory of Elasticity (2nd edn), 10—15 (Pergamon, Oxford, 1970)

[02] Lakes R S. *Science* 235, 1038—1040 (1987)

[03] Friis E A, Lakes R S & Park J B. *J. Mat. Sci.* 23, 4406—4414 (1988)

[04] Gibson L J & Ashby M F. Cellular Solids: Structure and Properties 70—82 (Pergamon, Oxford, 1988)

[05] Caddock B D & Evans K E. *J. Phys.* D 22, 1877—1882 (1989)

[06] Evans K E & Caddock B D. *J. Phys.* D 22, 1883—1887 (1989)

[07] Evans K E. *Chem. Ind.* 20, 654—657 (1990)

[08] Evans K E. *J. Phys.* D 22, 1870—1876 (1989)

[09] Lempriere B M. Am. Inst. Aeronaut. *Astronaut. J.* 6, 2226—2227 (1968)

[10] Alderson K L & Evans K E. *Polymer* (submitted)

[11] Gibson L J, Ashby M F, Schajer G S & Robertson C I. *Proc. R. Soc. Lond.* A382 25—42 (1982)

[12] DISCOVER, *Biosym Technologies Inc.*, 10065 Bames Canyon Road, San Diego, California 92121, USA

第20章 获奖演说
Chapter 20　Award-Receiving Speeches

20.1　The Discovery of Crown Ethers[①][②]

Ladies and Gentlemen, Dear Colleagues,

This is a wonderful day in my life, and I am looking forward to sharing my thoughts with you.

Before I begin, I would like to convey the warm greetings of the people of Salem County, New Jersey—where I have lived for many years—to the people of Sweden. Salem County is where a very early Swedish settlement was established in 1643. Next year we will join with the people of our neighboring state of Delaware to celebrate the 350th anniversary of the first landing of Swedes in the New World at the Rocks in Wilmington, Delaware. We look forward to the visit of his Majesty King Carl XVI Gustaf and her Majesty Queen Silvia and others from Sweden to our celebration next April.

Now I would like to discuss the discovery of the crown ether. I will divide my lecture into three parts.

First, because every discovery takes place in more than a scientific context, I would like to touch on my life and background. In the weeks since it was announced that I would share this year's prize in chemistry, people have expressed as much interest in my early life as they have in my later work. So I think it appropriate to express myself on the matter. It may also be that details of my past have more than casual bearing on my work.

Second, I would like to describe for you my research program and some of the specific events that led to the discovery of the first crown ether. Since I am the only one who knows at first hand the excitement and pleasure of the discovery, I will devote a portion of my time to sharing this experience with you.

And third, I would like to discuss the properties and preparation of crown ethers. In doing so, I hope I will convey to you that I was always a "hands-on" chemist; I took satisfaction from what I did in the laboratory. Also, I was very much an industrial chemist and was always interested in the potential application of my work. In fact, when I submitted my first major

---

①　Reprinted from: Nobel Lecture, December 9, 1987; Charles J Pedersen, E I. Du Pont de Nemours and Company, Wilmington, Delaware 19898, USA.

②　The remaining paragraphs and the references are omitted.

paper on the discovery of the crown ethers, the editor of the *Journal of the American Chemical Society*, Marshall Gates, remarked that my descriptions were replete with industrial jargon. Fortunately he published the paper anyway.

Personal Background

Let me start then with how I began life and went on to discover the crown ethers.

My father, Brede Pedersen, was born in Norway in 1865 and trained as a marine engineer. Due to sibling disharmony, he left home for good as a young man and shipped out as an engineer on a steam freighter to the Far East. He eventually arrived in Korea and joined the fleet of the Korean customs, which was administered by the British. He rose in rank and later joined one of the largest Japanese steamship lines and became a chief engineer. Then a tragedy occured that changed the course of his life. A childhood disease took the life of my elder brother while my father was away from home on a long journey. He abandoned the sea and became a mechanical engineer at the Unsan Mines in what is now the northwestern section of present-day North Korea.

My mother, Takino Yasui, was born in 1874 in Japan. She had accompanied her family to Korea when they decided to enter a large-scale trade in soybeans and silkworms. They established headquarters not far from the Unsan Mines, where she met my father.

The Unsan Mines were an American gold and lumber concession, 500 square miles in area. It had been granted by the Emperor of Korea to an American merchant named James R. Morse prior to 1870. I was conceived there in mid-winter just before the start of the Russo-Japanese war. Frequent incursions by Cossacks across the Yalu River into the region of the mines were considered to endanger my mother, so she and several American ladies were sent south by carriage to the railhead for safety. I was thus born on October 3, 1904, in the southern port of Fusan, the largest in Korea. My arrival was doubly welcomed because mother was still grieving the loss of her firstborn. She devoted the next 10 years to overseeing my education and that of my sister, Astrid, five years my senior, in foreign language schools.

I spent my first and last winter at the mines when I was 4 years old. The region was known for severe weather due to the confluence of the Siberian steppes, Mongolian Gobi Desert and the mountains of Korea. Large Siberian tigers still roamed the countryside and were frightened away with bells on the pony harnesses. Wolves killed children during the cold winter nights, and foxes slept on roofs against the chimneys to keep warm.

Because the Unsan Mines were an American enclave—the top management being all Americans—great emphasis was placed on making life as American as possible. The country club was the center of social activities and life was considerably more gentle than at the typical gold mine of the legendary American West. So my contacts with Americans began early, and I spoke English which was the common language at the mines.

I do not know if such an environment had a lifelong influence on me, but I can speculate that perhaps it did. Freedom of the Americans to administer their affairs in taking care of themselves in the wilds where things could not be ordered for overnight delivery no doubt taught a certain independent approach to problem solving. As for chemistry, I recall that the gold was recovered by the cyanide process, and the monthly cleanup day was marked by the pervasive odor of the process. The pouring of the molten gold was always a beautiful sight, and that might have started my interest in chemistry. Also, my sister claimed that I loved to play with a collection of colorful Siberian minerals.

Foreign languages schools did not exist in Korea then, and so at the age of 8, I was sent to a convent school in Nagasaki. When I was 10 years old, my mother took me to Yokohama where she remained with me for a year as I began my studies at St. Joseph College. St. Joseph was a preparatory school run by a Roman Catholic religious order of priests and brothers called the Society of Mary. There I received a general secondary education and took my first course in chemistry.

When it came time for me to start my higher education, there was no question of where it would be obtained. I had lived among Americans and had determined, with my father's encouragement, to study in America. I selected the University of Dayton in Ohio for two reasons: firstly, we had family friends in Ohio, and secondly, the same organization, the Society of Mary, ran both St. Joseph College and the University of Dayton.

My four years in Dayton and a year in graduate school at Massachusetts Institute of Technology were pleasant and taken up with activities that made me into an American. This perhaps also molded my scientific character and represented something of a personal metamorphosis. The sequence—Dayton first and then MIT—was also good, making a false start by a young man much less likely. The University of Dayton was a college of 400 men, most of them living in dormitories under strict monastic regimen. Training of the spirit was considered as important as training of the body and soul. I enjoyed all phases of the training. I became vice president of my graduating class, won letters in tennis and track and a gold medal for excellence that reflected my four years of performance there. Excellence in general was encouraged; I was even awarded a gold medal for conduct.

MIT was another matter. Boston, where I lived, is an old city of great charm and a center of the arts. I did not apply myself to my courses as I should, but my extracurricular activities contributed to the formation of my ultimate character. It was while studying at MIT that I first felt the exhilaration of utter freedom. MIT was considered deficient in the humanities, but with a little effort that deficiency could be remedied delightfully by visiting second-hand book stores. Why second-hand books appealed to me more than library books still remains a mystery—though it possibly was the prospect of finding unexpected treasures. I celebrated my graduation from MIT as a chemist by taking a walking tour of the Presidential Range in New Hampshire.

In spite of the urging of James F. Norris—a very prominent professor and my research advisor—I did not remain at MIT to take a Ph. D. My bills were still being paid by my father, and I was anxious to begin supporting myself. In 1927, I obtained employment at Du Pont through the good offices of Professor Norris, and I was fortunate enough to be directed to research at Jackson Laboratory by William S. Calcott. My career of 42 years had begun.

The research environment at Du Pont during those years was not altogether typical of industrial laboratories of the time. The company had formed the nucleus of a basic research department that in a few years' time would have scientist such as Wallace Carothers and the young Paul Flory working on the polymer studies that led to nylon and other breakthroughs. And in general, Du Pont was a productive center of research where many interesting and important problems were being solved. For example, one day while visiting Julian Hill at the Du Pont Experimental Station in Wilmington, Delaware, I observed him pull the first oriented fiber of a polyester. On another occasion, at Jackson Laboratory, across the Delaware River in New Jersey where I worked, I noticed commotion in the laboratory of Roy Plunkett, which was across the hall from my own. I investigated and witnessed the sawing open of a cylinder from which was obtained the first sample of Teflon® fluoropolymer. At Jackson Laboratory, during that time, other important advances were taking place in tetraethyl lead and new petroleum chemicals, new elastomers, and a new series of fluorocarbons for refrigeration and aerosols. The atmosphere was vibrant and exciting, and success was expected. It was in this atmosphere I began my career.

As a new scientist I was initially set to work on a series of typical problems, the successful solution of which buoyed my research career[01~05]. After a while, I began to search for oil-soluble precipitants for copper, and I found the first good metal deactivator for petroleum products[06~08]. As a result of this work, I developed a great interest in the effects of various ligands on the catalytic properties of copper and the transition elements generally, and I worked in that field for several years. I noticed a very unusual synergistic effect wherein a metal deactivator greatly increased the efficacy of antioxidants[09~10].

So more and more, I became interested in the oxidative degradation of the substrates themselves, particularly petroleum products and rubber. As my interests moved in that direction, I left off working on metal deactivators and coordination chemistry. By the mid-1940s, I was in full career, having established myself in the field of oxidative degradation and stabilization[11~13]. I was independent in terms of the problems I might choose and had achieved the highest non-management title then available to a scientist at Du Pont. During the 1940s and 1950s, my interests became more varied. For example, I became interested in the photochemistry of new phthalocyanine adducts and of quinoneimine dioxides. I found some polymerization initiators, discovered that ferrocene was a good antiknock agent for gasoline, and made some novel polymers[14~23].

20.2　Concept and Innovation in Polymer Science[①]

In the past century we have witnessed a prodigious increase in the use of polymeric materials. Polymers are now approaching a position of primacy among the several classes of materials on which people of the present world depend for the necessities of clothing, shelter, and transport, not to mention articles of convenience and the frills which have come to be taken for granted. They rival metals in their importance. Already their volume, measured in cubic metres produced per annum, exceeds that of metals, and the growth seems destined to continue. New compositions of matter, created by ingenious applications of the methods of synthetic chemistry, have set the pace of this remarkable proliferation of macromolecular materials, manifested both in their diversity and in their aggregate quantity.

Unlike communications and data processing technologies, where spectacular advances have been triggered by a specific "breakthrough" (the invention of the transistor in this instance) no single discovery or innovation is identifiable as responsible for the equally spectacular growth of polymer technology. In contrast to synthetic dyestuffs, electric power, chemotherapy and atomic energy, each the offspring of a new scientific concept or discovery which sparked the creation of an industry where none existed theretofore, the utilisation of polymers has antecedents that predate recorded history. Polymers of natural origin—cellulose of wood and textiles, proteins of wool and hides and, in parts of the world, natural rubber—have served the needs of man from times immemorial. Techniques for processing these materials are among the oldest technologies. They evolved over many centuries.

Conspicuous advances in the adaptation of naturally occurring polymeric materials to the needs of man took place long before the molecular nature of polymeric substances was established through the work of Staudinger and Carothers and their contemporaries in the 1920s and 1930s. Vulcanisation of rubber, discovered in 1839, greatly enhanced its usefulness and marked the beginning of a major industry. Chemical modifications of cellulose susceptible to moulding and to extrusion in the form of fibres and films were discovered and developed during the latter part of the 19th century. These and other achievements of that period occurred more or less concurrently with the emergence of

---

①　Reprinted from: Perkin Medal Lecture, *Chemistry and Industry*, 369~372, 21 May 1977; Paul J. Flory (Recipient of the Nobel Prize in Chemistry in 1974), Department of Chemistry, Stanford University, Stanford, CA 94305, USA.

modern chemistry. However, the innovators responsible for these developments may have been influenced more by the adventurous spirit of the age than by the enlightenment this new-born science afforded at that time. They made their contributions without benefit of the knowledge that the objects of their endeavors are polymers consisting of chain molecules of great length. Lacking a perception of the essential chemical constitution of cellulose and other polymers of natural origin, they were obliged to proceed empirically. Chemical knowledge at their disposal provided methods and procedures, but little more.

1. Early Achievements

The first synthetic polymers to reach commercialisation made their appearance in the same era, that is, during the early phase of chemistry as a science but prior to general acceptance of the "macromolecular hypothesis" on the nature of polymers. As it had long been known that rubber could be decomposed into isoprene, it was natural to attempt synthesis of rubber from isoprene and its analogues, notably butadiene. When this was achieved, the chemical formulae used to represent the first synthetic rubbers showed several monomers joined to form cyclic compounds having molecular weights not much greater than the non-polymeric compounds in the main stream of chemistry. Polystyrene and other synthetic curiosities were similarly regarded, as were proteins and cellulose. It is more a tribute than a detraction to the genius of Leo H. Baekeland that he pioneered the field of thermosetting resins empirically, around the turn of the century, with little aid from chemical theory and principles that only later would elucidate the constitution of the materials he invented.

These early achievements, briefly noted, were to be eclipsed by the far-reaching advances of the succeeding era, and the resulting profusion of new polymeric materials commencing in the 1930s. That rapid advances in polymer technology, and the concomitant growth of major polymer industries, occurred on the heels of the emergence of concepts seminal to the body of knowledge we call polymer science was not a fortuity. Indeed, these technological achievement could not have been realised without the benefit of the ideas and insights provided by this fledgling science. Several prominent examples are illustrative.

In the late 1920s Wallace H. Carothers set as the objective for himself and his co-workers the synthesis of substances whose polymeric constitution would be beyond dispute, using for this purpose well established methods of organic chemistry. If the properties of a substance thus prepared duplicated those of materials like rubber, cellulose, starch, proteins and vinyl polymers whose molecular nature was in contention, then by inference it would follow that the latter also are polymeric. It is especially noteworthy that the ensuing epochal investigations were undertaken for the purpose of exploring a scientific hypothesis on the constitution of substances now known to be polymeric, and not with the object of discovering new compositions of matter, or of establishing a new industry. The hypothesis Carothers set out to explore was fully confirmed and, incidental to the primary effort, a great deal was learned about condensation polymerisation and the substances that could be synthesised by such processes. Knowledge acquired in this area afforded the impetus that led to a whole new class of linear polymers, including nylon and polyesters.

The commercial exploitation of nylon and succeeding members of the broad class of condensation polymers was facilitated enormously by the investigations of Carothers and his co-workers, conducted with the aim of understanding polymerisation reactions and the molecular nature of the substances thus produced. The developmental work that followed could be conducted rationally on the basis of sound principles governing condensation polymerisation, thanks to the attention and effort devoted to reconnaissance of the scientific terrain at the outset. Whether the same achievements could have been realised by empirical exploration is doubtful at best. A vastly greater effort certainly would have been required to gain the same ends empirically, without the enlightenment of scientific understanding.

Most of the synthetic rubbers that are used today were pioneered in the early years of modern polymer science. Their development drew heavily upon the knowledge then becoming available concerning polymerisation mechanisms and polymer structure, and upon the theory of cross-linking processes that plague the production of processible rubber on the one hand but are so essential for full realisation of its potential properties in the final product of manufacture on the other.

The creation of a synthetic rubber industry within the span of several years during World War II is especially remarkable. The success of this unparalleled effort stands as a tribute principally to engineering skill, to managerial organisation and to applied research conducted under the exigencies of the times. Yet, as anyone who participated in that effort will recall, pure and applied scientists collaborated closely. Ideas were freely exchanged on critical aspects of polymer structure, on the mechanism of polymerisation and on the colloidal description of the medium in which the complex process of emulsion polymerisation is effected. Concepts urgently gathered to meet the challenge of a war-born emergency fortunately were well conceived; they proved to be valid and therefore useful. It would be difficult to estimate their impact on the course of this development of a synthetic rubber industry in unprecedented haste. Most will agree, I think, that the insights brought to bear from the science sector served an essential function in steering the development through the difficulties that invariably arise in an undertaking of such scope.

2. Stereoregular Polymers

A decade later stereoregular polymers, notably isotactic polypropylene, were discovered. They appeared quite accidentally in the course of exploratory investigations of processes for converting olefins to polymers—investigations conducted independently in several laboratories during the 1950s. While the discovery came unexpectedly, and not as the result of a planned effort to produce crystalline polymers having especially desirable properties, the structural feature giving rise to the physical characteristics of these polymers was quickly comprehended. Years earlier Staudinger surmised that vinyl polymers are amorphous, rather than crystalline, owing to the random positioning of their substituents on either side of the backbone of the chain. Others voiced the same opinion. From this it followed, conversely, that if the substituents could somehow be marshalled to fall into regularly assigned positions, then the resulting polymer probably would be crystalline and hence exhibit a set of properties quite different from its stereoirregular, amorphous isomer. Hence, when crystalline vinyl polymers made their unexpected appearance, first in the vinyl ethers prepared by Carl Schildknecht and in the polyolefins prepared by Giulio Natta and his collaborators and by others, the explanation was at hand. Research therefore proceeded rapidly and an important class of new materials for moulded articles, fibres and films became commonplace.

3. Heterophase Copolymers

As a further, more recent example, I would direct attention to the heterophase block copolymers which are endowed with unusual properties of durability and resistance to impact. Abstruse investigations on the thermodynamics of interactions of polymer molecules in solutions and in melts conducted years earlier had predicted and explained the intrinsic incompatibility between most pairs of chemically different polymers. If, however, hybrid "block" copolymers are prepared by connecting long chains of one monomer with long chains of the other, then gross separation of the two hostile constituents is impossible. Their proclivity for segregation nevertheless asserts itself through the formation of separate, small domains on a micro scale. These domains are necessarily of a size commensurate with the blocks of the respective monomers; that is, they are at the level of the molecular dimensions of the polymer.

This important development was supported by theory and concepts made available through basic research conducted a number of years earlier. Those responsible for this "idle" research could be faulted, perhaps, for failing to comprehend the practical implications of their investigations. Whatever their lack of foresight, it must be borne in mind that mortals proceed only a step at a time and are seldom omniscient in their outlook. The basic theory was the product of curiosity and zest for understanding. Practical applications were not of immediate concern. Moreover, they were concealed by several turns on the winding road ahead. On the other hand, the technological development was profoundly dependent on theory which fortunately had been brought into existence earlier.

4. Rigid-chain Polymers

In the present decade we have witnessed the appearance of polymeric materials that rival steel in strength. The comparison gains emphasis if account is taken of the sixfold advantage in density; for the same weight, the tensile strengths of these polymeric materials range up to five times that of steel. I refer to the aramide fibres of which "Kevlar" is the first to be commercialised. The aramide polymers that are exploited by this development consist of highly extended polymer chains of limited flexibility. When dissolved, these "stiff-chain" polymers form anisotropic phases, or liquid crystalline domains, in which the macromolecules are aligned relative to one another. It is this spontaneous generation of a high degree of order that appears to be responsible for the properties of fibres formed from these anisotropic solutions.

The theory that explains this spontaneous ordering was put forward more than 20 years ago. Experiments on α-helical polypeptides conducted by Robinson C. in the late 1950s demonstrated the occurrence of phase separation in solutions of highly extended macromolecules, as had been predicted by theory. But reduction to practice and demonstration of practical utility, in the vernacular of patents, is comparatively recent. The precise chronology of events aside, it is a valid assertion that this development would have been severely handicapped without a clear comprehension of the factors responsible for the unusual behaviour and properties manifested by such systems and of their morphological features at a molecular level. Achievements to date probably are only a beginning. Application of basic principles has opened an entirely new vista on the potentialities of polymeric materials in applications where high strength and durability over wide ranges of conditions are required.

5. Importance of Basic Research

The examples above, and others that could be cited as well, demonstrate the essential importance of concepts and the perceptions they afford in providing the knowledge and intellectual nourishment the creative mind requires for the exercise of its capacity for innovation. Significant inventions are not mere accidents (legal criteria for patentability to the contrary notwithstanding). This erroneous view is widely held, and it is one that the scientific and technological community, unfortunately, has done little to dispel. Happenstance usually plays a part, to be sure, but there is much more to invention than the popular notion of a bolt out of the blue. Knowledge in depth and in breadth are virtual prerequisites. Unless the mind is thoroughly charged beforehand, the proverbial spark of genius, if it should manifest itself, probably will find nothing to ignite.

There are of course well known examples in which the inventor relied on fallacious concepts to achieve his invention. These examples are, I believe, the exception rather than the rule. At our present level of technological sophistication, creative invention without a firm grasp of underlying principles becomes increasingly rare. Blind discoveries may continue to occur, but the betting odds are much in favour of the enlightened innovator, and these odds continue to increase.

It is precisely here that basic research makes its greatest contribution to the advance of technology and the enlargement of its potentialities. This assertion contradicts the prevalent view that the measure of practical value of basic research lies in its success in uncovering nuggets of truth that can be commercialised. The mystique of unanticipated discoveries resulting from basic research into the laws of nature is widely accepted, both by scientists and by the public at large, as the main justification for support of basic research. Examples from the past are cited as proof. Accepting this view, scientists are prone to emphasize possibilities for "spin-off" in their research proposals as justification for support.

Without denying the importance of salient discoveries that cannot be foreseen, I contend that basic research has a more pervasive mission in advancing knowledge, in providing incisive concepts and in sharpening insights. These are the ingredients that nurture enduring innovations of the broadest scope. They do not enjoy the visibility and the attention-arresting qualities of radical discoveries, but I submit that these less tangible contributions of basic research are of greater importance than the latter. This promises to be increasingly true in the future.

There is a further, closely related function of basic research, one to which I have already alluded in the examples above. I refer to its role in guiding the development of an innovative idea over the precarious path from its inception in the laboratory to successful commercialisation. Solution of the innumerable difficulties that invariably beset the implementation of an invention demands the best that technical and scientific knowledge can provide. To proceed blindly, without the support of the best that such knowledge can offer, is wasteful of effort at the very least, and may diminish the chances of success as well.

For reasons cited, the improvement and extension of basic knowledge through research must be accepted as utter necessities for sustenance of continued technological advance. Basic research must not be treated as a dispensable adjunct whose cultivation can be postponed at the pleasure of the profit margin—not if industrial technology is to advance, or even to meet its competition.

Let me disavow at this juncture any intimation that servicing the needs of applied science and technology is the only value of pure science. It is the one of concern here. Others are therefore ignored in this context.

The "how to do it" motive of applied research is being emphasised at the present time to the perilous neglect of the concern for "why" that motivates basic research. The trend is prevalent not only in industrial laboratories where long-term basic research and exploratory research has been trimmed to the bone or eliminated altogether; it is becoming the imperative in government agencies that support research, and it is increasingly in evidence within universities. Faculty members feel themselves obliged to justify their research in terms of biological applications, solution of problems of the environment or any of the numerous other problems that confront contemporary society. Often, these obligations are self-imposed. The lure of research funds and administrative pressures are factors as well.

Certainly, solution of urgent current problems, like those mentioned, commands the highest priorities. But successful solution of them will, in most instances, require enlargement of the science base in relevant areas. The urgencies of the present must not be allowed to foreclose the options that only extension of knowledge can make available.

Polymer science has not been spared the impact of attitudes concerning scientific research that currently are prevalent both within the scientific community and among the public at large. Too little effort is being devoted to fundamental research in polymer science, an area that is immediately relevant to both biology and industrial technology. This should be of vital concern to the chemical industry. It should be their concern not only to redress the balance of effort in their own laboratories, but also to promote adequate attention to this field in universities and in other public institutions that engage in scientific research.

6. Future Prospects

What can one venture to say concerning the future? To this end it is useful to attempt assessment of the present in the frame of reference of the past. It has been suggested by Mandelkern that we are entering an age of macromolecules, just as our forebears passed successively through a stone age, a bronze age, etc. The evidence is apparent.

No gift of clairvoyance is required to predict continuation of the trend of increasing reliance on polymers. The directions that future developments may take are more difficult to foretell. Certainly, we have by no means exhausted the possibilities of polymeric materials that display exceptional properties. Being composed of long chains of structural units, polymeric molecules admit of almost endless variations in molecular constitution—variations that are reflected in their chemical, physical and mechanical properties. As a class, polymers are singularly versatile in this respect.

This potentiality for diversity is exploited most remarkably by nature in the highly differentiated array of biological macromolecules that serve the structural, chemical and regulatory functions of living organisms. If one should fall into the error of concluding that polymer science is nearing the end of the road, even a cursory inspection of biopolymers should suffice to convince him that we have a long way to go. Emulation of nature in the design and synthesis of macromolecules presents a challenge of incomparable magnitude. Although this is a goal we may never reach, efforts to this end may nevertheless yield rich rewards.

Indulging further in the risky business of forecasting the future, I would venture the prediction that polymers designed for special purposes will assume greater promience, as opposed to the mass-produced, broad-spectrum polymers that have dominated the recent past. Polymers for medical applications are illustrative. Others that will withstand extremes of temperature promise to replace metals for special applications. We are already witnessing the birth of a new class of "ordered" polymers, mentioned earlier, which are composed of rigid molecular chains. The extraordinary strength and durability of fibres made from them add a new dimension to the applications of polymeric materials. The scientific basis for these achievements suggests broad applicability of the underlying principle.

Possibilities for new uses of polymeric materials appear to be vast both in magnitude and in their diversity. Exploitation of these possibilities will require vigorous pursuit of new knowledge and a renewed commitment of effort to this end. These are matters that should command the earnest attention and vigorous support of the chemical industry.

20.3 Scientific Research Moves towards the 21st Century[①]

Most scientists are specialists and I am no exception. This means that an address like this is fraught with a special danger in that to expound one's speciality leaves the audience behind, whereas to offer a scientific tour leaves the speaker behind. Have you shared with me the experience of reading a book which covers many fields of natural or technical or social science, which is well written and convincing, and then one comes to a chapter on an area where one is oneself an expert. Here disaster strikes, for the author of such well written convincing material, is an ignoramus and arrogant with it. Then all faith in the rest of the book evaporates. (Speaking as a specialist, quantum mechanics seems to have a peculiar ability to draw pontifical nonsense in such books.) This is the danger of a Presidential Address at the B. A.[②] and I will try hard to avoid it, but it is difficult to talk about the future and how current decisions affect the future without straying into regions where one is not an expect; but at least I can confess this when it happens. Thus I will organize this address by explaining some remarkable developments in the application of mathematics to science and technology in a very broad way, a way which will invade new territory for exact science. Then I will look at the future of science into the 21st century, trying very hard to distinguish unimpeachable facts from conjectures or prejudices.

Let me start then by talking about my own field of theoretical physics, on how it is changing and what it can be expected to do in the coming decades.

When I say theoretical physics, it would be better to say using mathematics in a situation where full prediction is possible. Mathematics entered science in those

---

① Reprinted from: Presidential Address to the B. A. given in Sheffield on 11 September 1989, *Sci. Publ. Affairs* 4, 131~139 (1990). Sir Sam Edwards, FRS (Recipient of the Boltzmann Medal of the IUPAP in 1995), Cavendish Laboratory (Department of Physics, University of Cambridge), Madingley Road, Cambridge CB3 0HE, UK.

② B. A. means the Bachelor of Arts degree and the graduation ceremony.

problems, which though profound, were clearly and simply described, i. e. , through astronomy where the solar system is the laboratory. Since Newton's time phenomenon after phenomenon has fallen to an accurate mathematical description and this goes far wider than physics, indeed over all technology and to the molecular and now biological sciences. Of course much of science and technology does not involve mathematics, but a precise quantification and prediction is the goal of science and technology and whereas an engineer often has to rely on experience for design, more and more everything relies on complete calculation and less on empirical experience. Until recently this process moved along well established paths. Experimental data generated or confirmed mathematical equations which could then be used to predict the behaviour of a material, or an engine, or more recently the nature of a molecule, or a natural phenomenon. The mathematics of a motor car can be expressed in terms of the laws of elasticity, of combustion, of aerodynamics. It takes a big computer to handle it, but one can in principle put a whole car into the computer, and certainly the components of a modern vehicle are computed out in full detail, not ignoring experience of course, but strongly supplementing it. It would be more complicated for me to explain the analogous position in chemistry, but it is there also. The basic tool in all this is the calculus. No one in the modern world can consider themselves an educated person without understanding two great concepts.

Firstly, the mathematical tool which allows one to describe change is the calculus. It is at its simplest when used for continuous change and, although discontinuity has been familiar to pure mathematics for a long time, this kind of mathematics has now invaded science making a much more complicated version of the calculus necessary.

The other great tool is entropy, the precise quantification of disorder which came into science via the need to understand steam engines, but now is found to be basic in studies of the environment and in the social sciences and the commercial world. The older aspects of these two branches of mathematics will never be superseded, but recently strange new ramifications are occurring which enormously strengthen our power of describing the world. A revolution is taking place which will affect the way science explains and predicts in the coming decades. There are three new ways of looking at things. The first involves fractals.

To illustrate recent changes, think about two experiments. In the first put a source of heat inside a block of material and measure the temperature. Heat flows in steadily at a point, and the temperature will rise outwards from the source as time goes by in a smooth predictable way, via the heat conduction equation. Now consider the block of material to be a hard jelly or gum and pump water into it at some point. The jelly will fracture and the water will make a kind of fern-like progress; for example, if the jelly is clear and the water is an ink, a feathery, fingering ink pattern will be seen. Now these phenomena are not new; lightening must be the oldest of all puzzling observations. What is new is that a decent mathematical description has come into being, and an astonishingly simple way to simulate such things on a computer. Imagine you, my audience, start to wander aimlessly about this hall, and I stand in the middle of the floor and grab anyone within arms length. Anyone I grab has the power themselves to grab anyone within their arms reach. The rest just keep milling around. Clearly a cluster will be created, and once it has reached any decent size, very few wanderers will reach people stuck near me, and so on out.

The pattern of grabbed people is precisely the fern like pattern one has in lightening (except that it is symmetric around the start point of course). This process is an easy one to put on a computer, so suppose I want to expel oil from an oil field by pumping water into the ground to displace the oil. The water pushes the oil, but the flow is unstable and the water fingers into the oil, until a finger reaches the oilwell outlet. By just letting randomly moving points in a computer stick to the "inlet" and then to those already stuck one can avoid the horrors of unstable equations and accurately describe the oil/water flow. The fern like or fractal structures as Mandelbrot has named them, are found extensively in nature, for example in the clouds and in coastlines and islands, and whereas the traditional methods of solving for smooth solutions has no bearing on such problem we now can solve things like this: car engine knock, the spread of epidemics, the failure of complex structures and many more examples. From now on a complex pattern or rather complex patternless form is no excuse for abandoning a quantitative study.

The next example I will discuss is that of chaos. This is the use of a familiar word for what is becoming a precise concept. Let me illustrate this by an obvious and extreme example. Suppose I throw a stick into a smoothly flowing river so that it eventually grounds on a bank. If I throw it in a slightly different direction, or start it at a slightly different point, it will ground at a slightly different position on the back. If the river is roaring down in spate, then obviously even a small difference in starting point can mean my stick being caught up in a different eddy in the water and ending in a quite different place. Of course it has found a different environment. But suppose I had some pendulums which were joined by elastic cords which lose energy in friction and viscosity. When started swinging they will eventually come to rest with the elastic cords pulling them well out of the horizontal. Starting them off in slightly different positions can mean that they end in totally different positions, i.e., if we think of a mapping of the initial positions to the final positions, there is nothing smooth at all. Tiny differences initially result in huge differences at the end. If you visualize this, it is quite in accord with intuition. What is new to that instead of just saying "this is a hopeless mess, let's study something else" people are now getting an understanding of the phenomenon. What is astonishing is that whereas until very recently, people believed that systems were in general quite predictable with a few awkward exceptions, it is now clear that although in the strict sense things are predictable, i.e., there is proper causality in nature, in fact in general they are not, and the standard problem of "here are the initial conditions, work out what happens" is the rarity, not the other way round. An immediate point is that people are beginning to understand just how predictable say the weather is, and that one can get a much clearer idea of the limits, because of the need of gathering of adequate information, or of computing power which have to be applied to predictions. For example if to predict next year's weather we need to know the position of every wave in the sea, we are not going to do it!

 The final example I want to discuss is frustration, again a well known concept now entering the world of precise science. This entered science through magnetism, but has quite universal applications in natural, social and technical sciences. Atoms in a magnetic material can be thought of as little bar magnets with one big difference. With a bar magnet north pole everywhere attracts a south pole and repels a north pole. With atoms it depends how far they are apart. In iron, north attracts north, south attracts south, so a (cold) lump of iron gets all the atoms pointing the same way and it is ferromagnetic. However, the spacing of the manganese atoms in say manganese fluoride is different and they behave just like ordinary bar magnets and the north and south poles arrange to be opposite to their neighbours, so that although everything is nicely ordered, there is no sign of external magnetism. Now what happens if we put a little of iron or manganese into liquid copper and freeze it into a dilute alloy. If the distance between iron atoms is as it is in iron they go parallel. If it is as in MnF_2 they go antiparallel. If they are just at random, which indeed they are, the atoms are frustrated. There is no simple way they can align to reach a minimum of energy, and there are many different arrangements, given the large number of atoms in the alloy, which are one as good as the other. There will be some true minimum of energy, but it will be the ultimate needle in the haystack. This magnetic system is called a "spin-glass" but is mirrored in an enormous number of problems. Consider this problem: a merchant ship picks up and puts down cargo in ports around the world. At each port it picks up for other ports and puts down from other ports. How does its owner make the most money: i.e., in what order should he make his calls. He is frustrated. If he goes to Singapore after Hong Kong he clearly is having to delay the cargo to Hawaii, and so on. It is the same problem as the magnet; and the mathematical methods invented for the magnet, work for the "travelling salesman". Physics and economics are in the same boat, as are many other problems ranging from computer design to the best way to wire and pipe a factory, and even to a possible model of the organisation of the brain.

 These examples I have given show that a revolution in quantitative science is taking place in the way disordered systems are no longer difficult or even incomprehensible, but are drawn into the net of predictability which started centuries ago in astronomy. Not only can we expect further advances, but we can expect to be able to say just how much we can predict, and what is unpredictable, however well we know the laws and however big our computers are.

I have dwelt so far on the growing ability to understand disordered systems, but I would now like to move to the opposite position and discuss matters where we have bedrock facts. I spent a term as Chief Scientific Advisor to the Department of Energy and remember with pride the tradition there, started by Lord Marshall, of publishing the full numbers, assembled by the Energy Technology Support Unit (ESTU), on which government policy is based... (intentionally omitted)

Thus it seems to me that there are a series of hard facts and a series of near hard facts that will guide our scientific society into the 21st century.

... (paragraphs intentionally omitted)

Given these points, let us finally turn to science. My remarks about my own speciality can be echoed all across science. Many things are happening, exciting both by the knowledge they give us, and by the applications which will flow from them. Applications really should look after themselves but what about basic science. The old theory in this country was that basic science was something done by people who taught at university and its magnitude was proportional to the number of students coming forward in any discipline. In other countries, a level was set which had nothing to do with the predilections of 14-year-olds, and institutes were set up to do the research, with universities in a minor position. It doesn't matter much which way you do it as long as it works, but it isn't working now. The reasons are complex, and I don't believe they are to be solved by any simplistic approach. Research has become much more expensive. Students are scarce and anyway the original equation was not very convincing. Also, and most importantly, much research has become interdisciplinary. To be, say, a skilled organic chemist you need only have a degree in organic chemistry. But then you may be faced with a problem involving elaborate apparatus, complex computer controls, complex calculations which need someone with the appropriate degree. I know this personally for my group in Cambridge needs complex molecules which physicists and material scientists do not have the skill to make.

These facts lead me to the proposition that scientific groups must be of a size that can afford and properly exploit world class equipment. Moreover in many cases these groups must be interdisciplinary. Now I know there are occasional gifted loners, and occasionally great research is done without much being spent on it, but my message in this address concerns numeracy. I want to see British scientists invited to conferences not because they are scholars who can make penetrating comments but because they have state of the art results to report. You will see therefore that I am a whole hearted supporter of the initiative of the Advisory Board to the Research Councils (ABRC) to organize Interdisciplinary Research Centres in Britain, and my only regret is that we are not planning many more of such centres and offering many more of our scientists the opportunities thus opened. I will confess myself green with envy when I see how our leading industrial groups recruit people to solve the problems, without having to ask: will he be able to give a course in my department? This viewpoint is incompatible with 50 top class research institutions; we can always argue for the money, but we just don't have the people. I sometimes wish we could be like Japan and call all our higher education institutes universities. Why not? Let us have 100 universities. Mr. Wilson made six overnight, surely Mrs Thatcher need not baulk at buying 50 new letter heads. At least then a lot of people would be made happy, and we really couldn't ask for 100 world class departments in everything.

Clearly this address is getting out of hand, so let me end on a lighter note. After I graduated, I thought I should see a bit of the world and joined the Institute for Advanced Study at Princeton. I was the youngest there; the oldest was Albert Einstein. The institute bus stopped outside his house, just for him, and sometimes he got on. So here is a claim to fame: I used to sit next to Einstein on a bus. I am sure he was well aware of his greatness, but he affected an interest in the common man, and I remember him trying to make conversation with the bus driver. This young man was reading a "comic" depicting a frightful green monster about to pounce on what for those days was a scantily clad beauty. The young man wanted to know what the beast was going to do to the beauty, not make conversation with one of the greatest intellects of all time. I did of course, but pushy young physicists were commonplace in Princeton, and Albert Einstein wanted to talk to the bus conductor.

Later in the day, but quite early since of course he was an old man at this time, Einstein would go home, and one day a person of wild appearance called, asking for him. Since the Director, J. Robert Oppenheimer, harboured "secrets" in his safe, Oppenheimer was assigned a G-man, so when the wild visitor asked for the "next best after Einstein" (a phrase which confirmed suspicions concerning his sanity) he was eagerly diverted to the G-man who, however, simply sent him packing. The police were informed that the wild man had left in the direction of the town. They soon rang back with the report of apprehending a wild looking man, but when the institute people went to identify, they found that the police had arrested the professor of pure mathematics who was naturally not very pleased.

From all this I should draw a moral, but the only thing I can think of is "all is not what it seems".

第四部分 (Part IV)

附 录
(Appendices)

附录A 单位、常数等实用资料
Appendix A Tables of Units, Constants and Other Useful Material

Table A.1 Fundamental Constants[a]

Symbol	Quantity	Value [Common Rounded-off Value]
a_0	Bohr radius	$5.2917706(44) \times 10^{-11}$ m [5.29]
c	velocity of light in vacuum	$2.99792458(12) \times 10^8$ m/s [3.00]
e	elementary charge	$1.6021892(46) \times 10^{-19}$ C [1.602]
F	Faraday's constant	$9.648456(27) \times 10^4$ C/mol [9.65]
g	gravitational acceleration	9.80665 m/s^2 [9.81]
h	Planck's constant	$6.626176(36) \times 10^{-34}$ J·s [6.626]
k	Boltzmann's constant	$1.380662(44) \times 10^{-23}$ J/K [1.381]
N_A, L	Avogadro's number	$6.022045(31) \times 10^{23}$/mol [6.022]
m_e	electron rest mass	$9.109534(47) \times 10^{-31}$ kg [9.11]
m_n	neutron rest mass	$1.6749543(86) \times 10^{-27}$ kg [1.675]
m_p	proton rest mass	$1.6726485(86) \times 10^{-27}$ kg [1.673]
R	molar gas constant	$8.31441(26)$ J/(K·mol) [8.314]
		$8.20568(26) \times 10^{-2}$ L·atm/(K·mol) [0.0821]
		$8.20568(26) \times 10^{-5}$ m^3·atm/(K·mol) [8.21]
R_∞	Rydberg's constant	$1.097373177(83) \times 10^7$/m [1.097]
		$2.179907(12) \times 10^{-18}$ J [2.18]
V_m	ideal gas molar volume	$22.41383(70)$ L/mol [22.4]
		$2.241383(70) \times 10^{-2}$ m^3/mol [0.0224]

[a] The numbers in parentheses represent the standard deviation in the last significant figures cited. Values based on *J. Phys. Chem. Ref. Data*, 2, 663~734 (1973).

Table A.2 Commonly Used Prefixes

Decimal Location	Prefix	Prefix Symbol
10^{12}	tera	T
10^9	giga	G
10^6	mega	M
10^3	kilo	k
10^2	hecto	h
10^1	deca	da
10^{-1}	deci	d
10^{-2}	centi	c
10^{-3}	milli	m
10^{-6}	micro	μ
10^{-9}	nano	n
10^{-12}	pico	p
10^{-15}	femto	f
10^{-18}	atto	a

Table A.3 Base Units of the International System of Units[a]

Physical Quantity Unit	Abbreviation	Selected Conversion Factors [Common Rounded-off Value]
Length		
meter	m	
angström	Å	1 Å = 1×10^{-10} m = 0.1 nm*
inch	inch	1 inch = 2.54×10^{-2} m*
Mass		
kilogram	kg	
atomic mass unit	u	1 u = $1.6605655(86) \times 10^{-27}$ kg [1.6606]
metric ton		1 metric ton = 1×10^3 kg*
pound（英制）	lb	1 lb = 4.5359237×10^{-1} kg* [0.454]
ounce（英制）	oz	1 oz = $2.8349523125 \times 10^{-2}$ kg* [0.0283]
ounce（英国金衡制）	oz	1 oz = $3.11034768 \times 10^{-2}$ kg* [0.0311]
Time		
second	s	
Electrical current		
ampere	A	
Temperature		
Kelvin	K	
degree Celsius	°C	$T/K = T/°C + 273.15$*
degree Fahrenheit	°F	$T/°C = [T/°F + (40 \text{ F degrees})] \times (5/9) - (40 \text{ C degrees})$*
Amount of substance mole		
mole	mol	
Luminous intensity		
candela	cd	

[a] The SI units are shown in boldface. Selected conversion factors for non-SI units are given. The numbers in parentheses represent the standard deviation in the last significant figures cited. Factors marked with an asterisk are exact. Values based on *J. Phys. Chem. Ref. Data*, 2, 663~734 (1973).

Table A.4 Derived Units of the International System of Units[a]

Physical Quantity Unit	Abbreviation	Selected Conversion Factors [Common Rounded-off Value]
Area		
square meter	m²	
Density		
kilogram per cubic meter	kg/m³	
grams per milliliter	g/mL = g/cm³	1 g/cm³ = 1×10^3 kg/m³*
Dipole moment		
Coulomb meter	C·m	
Debye	D	1 D = $3.335641(14) \times 10^{-30}$ C·m [3.34]

续 表

Physical Quantity Unit	Abbreviation	Selected Conversion Factors [Common Rounded-off Value]
Electrical resistance		
Ohm	$\Omega = V/A = kg \cdot m^2/(A^2 \cdot s^3)$	
Electricity quantity		
Coulomb	$C = A \cdot s$	
electrostatic unit	$esu = cm^{3/2} \cdot g^{1/2}/s$	$1\ C = 3.335641(14) \times 10^{-10}\ esu\ [3.34]$
Electromotive force		
Volt	$V = kg \cdot m^2/(A \cdot s^3)$	
Energy		
Joule	$J = kg \cdot m^2 \cdot s^{-2}$	
erg	$erg = g \cdot cm^2/s^2$	$1\ erg = 1 \times 10^{-7}\ J^*$
calorie	cal	$1\ cal = 4.184\ J^*$
electron volt	eV	$1\ eV = 1.6021892(46) \times 10^{-19}\ J\ [1.602]$
liter atmosphere	$L \cdot atm$	$1\ L \cdot atm = 1.01325 \times 10^2\ J^*$
wave number	cm^{-1}	$1\ cm^{-1} = 1.986477(10) \times 10^{-23}\ J\ [1.986]$
atomic mass unit	u	$1\ u = 1.492442(6) \times 10^{-10}\ J\ [1.492]$
Entropy		
Joule per Kelvin	J/K	
adsorption unit, Gibbs	cal/K	$1\ cal/K = 4.184\ J/K^*$
Force		
Newton	$N = kg \cdot m/s^2$	
dyne	$dyn = g \cdot cm/s^2$	$1\ dyn = 1 \times 10^{-5}\ N^*$
pound force	lbf	$1\ lbf = 4.4482216152605\ N^*\ [4.45]$
Frequency		
Hertz	$Hz = s^{-1}$	
cycles per second	cps	$1\ cps = 1\ Hz^*$
Pressure		
Pascal	$Pa = N/m^2 = kg/(m \cdot s^2)$	
atmosphere	atm	$1\ atm = 1.01325 \times 10^5\ Pa^*$
bar	bar	$1\ bar = 1 \times 10^5\ Pa^*$
pounds per square inch	$psi = lb/in^2$	$1\ psi = 6.894757293167 \times 10^3\ Pa\ [6.89]$
Torr, millimeter of mercury	$Torr = mmHg$	$1\ Torr = 1.33322368421 \times 10^2\ Pa\ [133.3]$
		$1\ atm = 760\ Torr^*$
Radiation activity		
Becquerel	$Bq = disintegration/s$	
Rutherford	Rd	$1\ Rd = 1 \times 10^6\ Bq^*$
Curie	Ci	$1\ Ci = 3.7 \times 10^{10}\ Bq^*$
Radiation dosimetry		
Coulomb per kilogram	C/kg	
Roentgen	R	$1\ R = 2.57976 \times 10^{-4}\ C/kg^*\ [2.58]$

Physical Quantity Unit	Abbreviation	Selected Conversion Factors [Common Rounded-off Value]
Radiation, energy absorbed		
Gray	Gy = J/kg	
rad	rad	1 rad = 1×10^{-2} Gy*
Volume		
cubic meter	m^3	
liter	L	1 L = 1×10^{-3} m^3*
quart(美制液体计量单位)	qt	1 qt = 9.4635295×10^{-4} m^3* [9.46]
milliliter	mL	1 mL = 1cm^3*
cubic centimeter	cm^3 = cc	1 cm^3 = 1×10^{-6} m^3*

^a The SI units are shown in boldface. Selected conversion factors for non-SI units are given. The numbers in parentheses represent the standard deviation in the last significant figures cited. Factors marked with an asterisk are exact. Values based on *J. Phys. Chem. Ref. Data*, 2, 663~734 (1973).

Table A.5 Atomic Masses Listed Alphabetically

Element		Atomic number	Atomic mass
Name	Symbol		
Actinium	Ac(锕)	89	227.0278
Aluminum	Al(铝)	13	26.98154
Americium	Am(镅)	95	(243)
Antimony	Sb(锑)	51	121.75
Argon	Ar(氩)	18	39.948
Arsenic	As(砷)	33	74.9216
Astatine	At(砹)	85	(210)
Barium	Ba(钡)	56	137.33
Berkelium	Bk(锫)	97	(247)
Beryllium	Be(铍)	4	9.01218
Bismuth	Bi(铋)	83	208.9804
Boron	B(硼)	5	10.81
Bromine	Br(溴)	35	79.904
Cadmium	Cd(镉)	48	112.41
Calcium	Ca(钙)	20	40.08
Californium	Cf(锎)	98	(251)
Carbon	C(碳)	6	12.011
Cerium	Ce(铈)	58	140.12
Cesium	Cs(铯)	55	132.9054
Chlorine	Cl(氯)	17	35.453
Chromium	Cr(铬)	24	51.996
Cobalt	Co(钴)	27	58.9332
Copper	Cu(铜)	29	63.546
Curium	Cm(锔)	96	(247)

续 表

Element		Atomic number	Atomic mass
Name	Symbol		
Dysprosium	Dy(镝)	66	162.50
Einsteinium	Es(锿)	99	(252)
Erbium	Er(铒)	68	167.26
Europium	Eu(铕)	63	151.96
Fermium	Fm(镄)	100	(257)
Fluorine	F(氟)	9	18.998403
Francium	Fr(钫)	87	(223)
Gadolinium	Gd(钆)	64	157.25
Gallium	Ga(镓)	31	69.72
Germanium	Ge(锗)	32	72.59
Gold	Au(金)	79	196.9665
Hafnium	Hf(铪)	72	178.49
Helium	He(氦)	2	4.00260
Holmium	Ho(钬)	67	164.9304
Hydrogen	H(氢)	1	1.0079
Indium	In(铟)	49	114.82
Iodine	I(碘)	53	126.9045
Iridium	Ir(铱)	77	192.22
Iron	Fe(铁)	26	55.847
Krypton	Kr(氪)	36	83.80
Lanthanum	La(镧)	57	138.9055
Lawrencium	Lr(铹)	103	(260)
Lead	Pb(铅)	82	207.2
Lithium	Li(锂)	3	6.941
Lutetium	Lu(镥)	71	174.967
Magnesium	Mg(镁)	12	24.305
Manganese	Mn(锰)	25	54.9380
Mendelevium	Md(钔)	101	(258)
Mercury	Hg(汞)	80	200.59
Molybdenum	Mo(钼)	42	95.94
Neodymium	Nd(钕)	60	144.24
Neon	Ne(氖)	10	20.179
Neptunium	Np(镎)	93	237.0482
Nickel	Ni(镍)	28	58.69
Niobium	Nb(铌)	41	92.9064
Nitrogen	N(氮)	7	14.0067
Nobelium	No(锘)	102	(259)
Osmium	Os(锇)	76	190.2
Oxygen	O(氧)	8	15.9994
Palladium	Pd(钯)	46	106.42
Phosphorus	P(磷)	15	30.97376

续 表

Element		Atomic number	Atomic mass
Name	Symbol		
Platinum	Pt(铂)	78	195.09
Plutonium	Pu(钚)	94	(244)
Polonium	Po(钋)	84	(209)
Potassium	K(钾)	19	39.0983
Praseodymium	Pr(镨)	59	140.9077
Promethium	Pm(钷)	61	(145)
Protactinium	Pa(镤)	91	231.0359
Radium	Ra(镭)	88	226.0254
Radon	Rn(氡)	86	(222)
Rhenium	Re(铼)	75	186.207
Rhodium	Rh(铑)	45	102.9055
Rubidium	Rb(铷)	37	85.4678
Ruthenium	Ru(钌)	44	101.07
Samarium	Sm(钐)	62	150.36
Scandium	Sc(钪)	21	44.9559
Selenium	Se(硒)	34	78.96
Silicon	Si(硅)	14	28.0855
Silver	Ag(银)	47	107.868
Sodium	Na(钠)	11	22.98977
Strontium	Sr(锶)	38	87.62
Sulfur	S(硫)	16	32.06
Tantalum	Ta(钽)	73	108.979
Technetium	Tc(锝)	43	(98)
Tellurium	Te(碲)	52	127.60
Terbium	Tb(铽)	65	158.9254
Thallium	Tl(铊)	81	204.383
Thorium	Th(钍)	90	232.0381
Thulium	Tm(铥)	69	168.9342
Tin	Sn(锡)	50	118.69
Titanium	Ti(钛)	22	47.88
Tungsten	W(钨)	74	183.85
Unnilhexium	Unq(?)	106	(263)
Unnilpentium	Unp(?)	105	(262)
Unniquadium	Unh(?)	104	(261)
Uranium	U(铀)	92	238.0289
Vanadium	V(钒)	23	50.9415
Xenon	Xe(氙)	54	131.29
Ytterbium	Yb(镱)	70	173.04
Yttrium	Y(钇)	39	88.9059
Zinc	Zn(锌)	30	65.38
Zirconium	Zr(锆)	40	91.22

附录 A 单位、常数等实用资料(Tables of Units, Constants and Other Useful Material)

Table A.6 Greek Alphabet

Name	Symbol Lower case	Symbol Upper case
alpha	α	A
beta	β	B
gamma	γ	Γ
delta	δ	Δ
epsilon	ε	E
zeta	ζ	Z
eta	η	H
theta	θ	Θ
iota	ι	I
kappa	κ	K
lambda	λ	Λ
mu	μ	M
nu	ν	N
xi	ξ	Ξ
omicron	o	O
pi	π	Π
rho	ρ	P
sigma	σ	Σ
tau	τ	T
upsilon	υ	Υ
phi	φ	Φ
chi	χ	X
psi	ψ	Ψ
omega	ω	Ω

Table A.7 List of Audio-Video Material Teaching Basic Chemistry

1. *The Super-charged World of Chemistry* (Videotype & DVD, Standard Deviants Video Course Review), Debbie Mintz, Laura Durrett, Kristie Wingenbach, David Rowley, Chris Fetner, et al; Cerebellum Corp. , 1996.

   提要：The Standard Deviants take on the Complex World of Chemistry. Pt. 1 shows the eternal struggle between superheroes and villains while reviewing difficult concepts such as percent composition and stoichiometry. Pt. 2 reviews thermochemistry, atomic structure, and chemical bonding. Pt. 3 reviews VSEPR theory, kinetic molecular theory and Dalton's law of partial pressures.

2. *Chemistry TV*: *Choices for Organic Chemistry* (书号: 0763709727) & *Core Organic Chemistry*, Version 2.0.1 *Faculty edition* (书号: 0763705926), Luceigh B A, Jones & Bartlett Pub (1997, CD-ROM).

3. 《化学专业英语多媒体教学课件》(出版中)，许昌学院"化学专业英语课件"课题组，2012.

4. 其他录像或光盘举例：

   (1) Origin of the Elements; (2) Designing Molecules; (3) Elements Organized: the Periodic Table; (4) The Halogens; (5) Chemical Families; (6) Chemical Lab Safety; (7) Chemical Bond & Atomic Structure; (8) Deterioration of Water; (9) 化学九个节目：碳及其化合物；化学键；设计；分子；物质三态转变；水变质；化学键和原子结构；扩散和渗透；光合作用。

附录 B 补充习题与已有习题答案
Appendix B Additional Homework and Answers to Existing Homework

B.1 Additional Homework

Additional Homework No. 1

01 In your own words, write the definitions or explanations for the following significant terms:

(01) periodic table;   (02) diffraction;   (03) atomic orbital;
(04) period;   (05) Lewis symbol;   (06) ionic bonding;
(07) oxidation state;   (08) structural isomer;   (09) hydrogen bond;
(10) chemical equilibrium;   (11) ligand;   (12) bond order;
(13) complex;   (14) enantiomer;   (15) synfuel;
(16) nucleophile;   (17) sublimation;   (18) space lattice;
(19) non-stoichiometric compound;   (20) miscibility.

02 From the following definitions or explanations, write down the corresponding significant terms.

(01) any property of a solution that depends on the relative numbers of solute and solvent particles.

(02) the passage of solvent molecules through a semipermeable membrane from a more dilute solution into a more concentrated solution.

(03) the short-lived combination of reacting atoms, molecules, or ions that is intermediate between reactants and products.

(04) substances that make a catalyst more effective.

(05) which is qual to the product of the concentrations of the reaction products, each raised to the power equal to its stoichiometric coefficient, divided by the product of the concentrations of the reactants, each raised to the power equal to its stoichiometric coefficient.

(06) a measure of disorder.

(07) the oxidized and reduced species that appear in an ion-electron equation.

(08) the process of driving a nonspontaneous redox reaction to occur by means of electrical energy.

(09) a measure of the potential difference between the two half-cells.

(10) which is concerned with the problems attending the determination of the amount of species present in a given sample.

(11) which relies upon optical, electrochemical, and other physical or physicochemical properties of sample solutions.

(12) the method where an electrical property is determined.

(13) that value about which all others are equally distributed, half being numerically greater and half being numerically smaller.

(14) which is equal to the square root of the quantity obtained by the division of the sum of the squares of absolute deviations by the number of times of measurements minus one.

(15) the division of the absolute error by the accepted value.

(16) a direct comparison of each weight in a set with one whose value is known with certainty.

(17) a process wherein the capacity of a substance to combine with a reagent is quantitatively measured. Ordinarily this is accomplished by the controlled addition of a reagent of known concentration to a solution of the substance until reaction between the two is judged to be complete; the volume of reagent is then measured.

(18) which enable the analyst to deliver any volume up to the maximum capacity.

(19) which in different regions of the electromagnetic spectrum gives different kinds of information about the structure and geometry of molecules.

(20) the distribution of a solute between a stationary and a mobile phase. It allows substances to be separated on the basis of their relative affinities for the two phases. Generally the stationary phase is a solid or a liquid, while the moving phase is a liquid or a gas.

03 Give Chinese names for the following significant terms:

(01) thermoplastic polymer; (02) copolymer; (03) glass transition temperature;
(04) vulcanization; (05) polyamide; (06) poly(vinyl chloride) or PVC;
(07) nylon; (08) isotactic polymer; (09) interpenetrating polymer network;
(10) intrinsic viscosity; (11) carbohydrate; (12) lipid;
(13) protein; (14) nucleic acids; (15) DNA;
(16) anabolism; (17) collagen; (18) keratin;
(19) primary structure; (20) quaternary structure.

Additional Homework No. 2

01 Answer the 4 questions in Section 18.4 on p. 156 of this textbook.

02 Write a lab report answering all the 5 questions in Section 18.5 on pp. 156~159 of this textbook.

03 Give an estimate of the values of chemical shifts for the 4 types of protons in 3-methyl-1-butene on p. 160.

04 Describe some applications of computational chemistry.

05 Write down the structural formula for the enol acetate as the C-(5, 6) dibromide in Section 19.1 on p. 164 of this textbook.

06 What is the main difference between Gaussian model and Edwards model? Draw a picture of an endlooped random walk or macromolecule.

07 What does a speech differ from a paper in?

Additional Homework No. 3

01 Do the Ex. 02 (abstract-writing) in "Comprehension and Writing" of C.3 on p. 210 of this textbook.

02 Answer the 14 questions in the COMPREHENSION section of Reading (03) in D.1 on p. 218 of this textbook.

03 Translate into Chinese the article of Translating (02) in D.2 on pp. 223~224 of this textbook.

04 Translate into English the first and last paragraphs of D.3 on pp. 224~225 of this textbook.

05 Translate into Chinese all the terms of Appendix F on pp. 249~257 of this textbook.

B.2 Answers to Existing Homework

Homework No. 2

01 (a) 500; (b) 68.46; (c) 10.0; (d) 11000; (e) 2.3; (f) 19.2; (g) 900; (h) 850; (i) 170; (j) 42.2; (k) 18.0; (l) 127(127.4558).

02 (a) 2.43; (b) 2.43±0.07; (c) 0.4‰; (d) No

03 (a) 6.066×10^3; (b) 1.64×10^{-14}; (c) 5.14×10^{-8}; (d) 1.366×10^3; (e) 1.00; (f) 70 km

04 (a) $x°C = 5(x°F + 40)/9 - 40 \Rightarrow -40°C$, i.e. $-40°F$;
(b) 160°C, i.e. 320°F;
(c) $-80/7°C = 80/7°F = 11.4°F$

05 (a) 24 Å^2; (b) 1.4×10^5 m^2

06 (a) 8.314×10^7 erg/(K·mol); (b) 1.987 cal/(K·mol); (c) 0.08205 L·atm/(K·mol)

07 (a) 25.052 cm^3; (b) 21.073 g; (c) 17.601 g; (d) 17.601 cm^3; (e) 7.451 cm^3; (f) 2.828 g/cm^3

08	$^A_Z\text{Element}^{n\pm}$	$(A=Z+N)$				
	$^{190}_{78}\text{Pt}$	78	112	190	78	0
	$^{139}_{53}\text{I}^-$	53	86	139	54	-1
	$^{29}_{14}\text{Si}$	14	15	29	14	0
	$^{188}_{79}\text{Au}^{3+}$	79	109	188	76	$+3$

09   (a) $^{12}_7\text{N}, ^{13}_7\text{N}, ^{14}_7\text{N}, ^{15}_7\text{N}, ^{16}_7\text{N}, ^{17}_7\text{N}$

      (b) $^{13}_5\text{B}, ^{14}_6\text{C}, ^{15}_7\text{N}, ^{16}_8\text{O}, ^{17}_9\text{F}, ^{18}_{10}\text{Ne}$

      (c) $^{13}_5\text{B}, ^{13}_7\text{N}$; $^{14}_6\text{C}, ^{14}_7\text{N}$; $^{16}_8\text{O}, ^{16}_7\text{N}$; $^{17}_9\text{F}, ^{17}_7\text{N}$

10   ^{35}Cl: 75.8%, 5.8069×10^{-23} g

     ^{37}Cl: 24.2%, 6.1388×10^{-23} g

Homework No. 3

01   (a) Na^+, Zn^{2+}, Ag^+, Hg^{2+}, Fe^{3+}, Fe^{3+}; Li^+, Bi^{3+}, Fe^{2+}, Cr^{3+}, K^+; P^{3-}, S^{2-}, Te^{2-}, Cl^-, I^-

      (b) NaF, ZnO, BaO_2, MgBr_2, HI, NaN_3, Ca_3P_2, FeO, AgF, CuCl, KN_3, MnO_2, Fe_2O_3

      (c) K_2SO_3, $\text{Ca}(\text{MnO}_4)_2$, $\text{Ba}_3(\text{PO}_4)_2$, Cu_2SO_4, $\text{NH}_4(\text{CH}_3\text{COO})$, $\text{Fe}(\text{ClO}_4)_2$, KNO_2, Na_2O_2, $(\text{NH}_4)_2\text{Cr}_2\text{O}_7$, Na_2CO_3; AgNO_3, $\text{U}(\text{SO}_4)_2$, $\text{Al}(\text{CH}_3\text{COO})_3$, $\text{Mn}_3(\text{PO}_4)_2$

      (d) B_2O_3, SiO_2, PCl_3, SCl_4, BrF_3; IBr, N_2S_5, PI_3, SiS, S_4N_2

02   (a) $\text{Al}_2\text{S}_3\,(s) + 6\text{H}_2\text{O}\,(l) \longrightarrow 2\text{Al(OH)}_3\,(s) + 3\text{H}_2\text{S}\,(g)$

      (b) $\text{O}_3\,(g) + \text{NO}\,(g) \longrightarrow \text{NO}_2\,(g) + \text{O}_2\,(g)$

      (a) Solid silicon tetraiodide reacts with liquid water in the presence of excess water to produce solid silicon dioxide and an aqueous solution of hydrogen iodide or hydroiodic acid

      (b) Aqueous arsenous acid reacts with gaseous hydrogen sulfide to give solid arsenic (Ⅲ) sulfide and liquid water

03   (a) 98.00 g/mol; (b) 193.07 g/mol; (c) 366.09 g/mol; (d) 360.39 g/mol

04   (a) 2.2×10^{23} atoms; (b) 2.2×10^{23} molecules; (c) 2.2×10^{23} formula units

05   171.07 g; 68.43 g; 20.13 g

06   (a) 0.0915 mol; (b) 0.0915 mol; (c) 22.8 g

07   0.067 mol; 0.067 mol; 0.22 mol/L

08   (a) Y: 44.9308%, Al: 22.7263%, O: 32.3429%; (b) 0.18 g

09   51.9 g/mol, Cr (51.996 g/mol)

10   C: 0.52 g = 0.043 mol, O: 2.43 g, H: 0.13 g = 0.13 mol;

     O in ROH: 0.35 g = 0.021875 mol;

     O: 1, C: 0.043/0.021875=1.9657, H: 0.13/0.021875=5.9428 $\Longrightarrow$ $\text{C}_2\text{H}_6\text{O} = \text{C}_2\text{H}_5\text{OH}$

Homework No. 4

01   (a) 2 mL; (b) Dalton's law; (c) the volume decreases; (d) the volume increases;

      (e) $\text{XeF}_4 < \text{Xe} < \text{F}_2$; (f) XeF_4

02   $-78.5\,°C$, 6.524 L; $-195.8\,°C$, 2.594 L; $-268.9\,°C$, 0.143 L; ca. $-273.15\,°C$

03   410 K

04   PCl_5

05   2.19×10^8 L $= 2.19 \times 10^5$ m^3; 17.86 m

06   $M = 52.1$ g/mol; 46.2/12=3.85, 53.8/14=3.84 $\Longrightarrow$ CN; $(\text{CN})_2$

07   $M = 2.858 \times 0.0820568 \times 273.13 = 64.06$ (SO_2: 64 g/mol)

08   (a) 28.962 g/mol; (b) 28.9516 g/mol; (c) nearly the same; (d) air is a mixture; (e) 1.16 g/L;

      (f) 1.15 g/L; (g) H_2O is less dense than the air it replaces; (h) 0.0003 atm; (i) 8.2556 g;

(j) 10.10 atm; (k) 9.11 atm; (i) 10.86%

Homework No. 5

01 (a) $Fe_3O_4(s) + 4H_2O(g) \longrightarrow 3Fe(s) + 4H_2O(l)$: type (iii) reaction

(b) $2KClO_3(s) \longrightarrow 2KCl(s) + 3O_2(g)$: type (ii) reaction

(c) $C(s) + H_2O(g) \xrightarrow{\Delta} CO(g) + H_2(g)$: type (iii) reaction

(d) $Cl_2O_7(g) + H_2O(l) \longrightarrow 2HClO_4(aq)$: type (i) reaction

(e) $Br_2(l) + H_2O(l) \longrightarrow HBr(aq) + HBrO(aq)$: type (iv) reaction

(f) $Ca_3(PO_4)_2(s) + 3H_2SO_4(aq) \longrightarrow 3CaSO_4(s) + 2H_3PO_4(aq)$: type (iv) reaction

(g) $2K(s) + 2H_2O(l) \longrightarrow 2KOH(aq) + H_2(g)$: type (iii) reaction

(h) $MgCO_3(s) \longrightarrow MgO(s) + CO_2(g)$: type (ii) reaction

02 (a) No; (b) Yes; (c) No; (d) Yes; (e) No; (f) No; (g) Yes; (h) No

03 (a) $PbSO_4$; (b) $Na^+ + CH_3COO^-$; (c) $2NH_4^+ + CO_3^{2-}$; (d) MnS; (e) $Ba^{2+} + 2Cl^-$;

(f) $2NH_4^+ + SO_4^{2-}$; (g) $Na^+ + Br^-$; (h) $Ba(CN)_2$; (i) $Mg(OH)_2$; (j) $2Li^+ + CO_3^{2-}$

04 (a) (i) CO_2, H_2O; (ii) $C_6H_{12}O_6$, O_2; (iii) chlorophyll; (b) light;

(c) $6CO_2(g) + 6H_2O(l) \xrightarrow[\text{chlorophyll}]{\text{light}} C_6H_{12}O_6(aq) + 6O_2(g)$;

(d) (i) 1:1; (ii) 1:6; (iii) 1:1;

(e) Yes;

(f) No, because $\Delta n = 0 \Rightarrow V \cdot \Delta p = 0$

05 $O_2 : N_2 = 1 : 2$; Air: $O_2 : N_2 = 21 : 78 = 1 : 3.7 \Rightarrow$ Yes

06 $d = 0.00185$ cm (on each side)

07 (a) $4HNO_3(aq) + Cu(s) \longrightarrow Cu(NO_3)_2(aq) + 2NO_2(g) + 2H_2O(l)$

(b) 0.8488 mol of Cu and 0.212 mol of HNO_3 (limiting reactant)

(c) 2.56 L

(d) Nothing would happen

08 (a) 7.05 g; (b) 95.0%

09 $FeS_2(s) \longrightarrow 2SO_3(s) \longrightarrow 2H_2SO_4 \Rightarrow 1.63$ kg

Homework No. 6

01 (a) Work is negligible; (b) & (c) Work is done by the surroundings on the system

02 $\Delta H_{sub} = \Delta H_{vap} + \Delta H_{fusion}$

03 (a) -12.4 kJ; (b) 4.96 kJ

04 (a) 10.519 kJ/mol; (b) 310.114 kJ/mol; (c) 320.633 kJ/mol

05 -2.98×10^4 kJ; -9.28 kJ

06 Al: 24.3; Be: 16.416; Cr: 23.92; Fe: 25.24; Pb: 26.936; Sn: 26.8

07 0.523 J/(K·g); 49.7 g/mol $\Rightarrow$ V [50.9415 g/mol]

08 (a) 0.13 J/(K·g) $\Rightarrow$ W; (b) 0.266 J/(K·g) $\Rightarrow$ Mo

09 (a) -6.335 kJ; (b) 5.882 kJ; (c) -92.30 kJ; -92.76 kJ, More exothermic

Homework No. 7

01 IUPAC = International Union of Pure and Applied Chemistry; -ane; By using prefix like metha-, etha-, propa-, buta-, penta-, hexa-, hepta-, octa-, and deca-; No, the correct one is 2,2-dimethylpentane.

02 $CH_3-CH_2-CH_2-CH_2-CH_3$, n-pentane; $CH_3-CH_2-CH(CH_3)_2$, 2-methylbutane; $C(CH_3)_4$, 2,2-dimethylpropane with one NMR peak.

03 An alkyl group is one that contains one less hydrogen atom than alkane;

C_nH_{2n+1} and C_nH_{2n-1}; Replacing the suffix -ane by -yl in an alkane.

04 Location of the double bond is unclear.

05 (a) C—C—C=C—C—C; (b) C—C=C—C=C; (c) □
(d) C—C—CH(CH$_2$CH$_3$)—CH(CH$_2$CH$_3$)$_2$; (e) C—C—C=C; (f) C—CH(CH$_3$)=C;
(g) C—CH(CH$_3$)—C(CH$_2$CH$_3$)=C; (h) C—CH(CH$_3$)—C=C; (i) p-Ph(NO$_2$)$_2$;
(j) Ph—CH$_2$CH$_2$CH$_3$; (k) m-Ph(Br)$_3$; (l) C—C(Ph)—C(Ph)—C.

06 *cis*- or *trans*-; If the methyl group and the propyl group are on the same side of the double bond, the name should be *cis*-2-hexene, otherwise, *trans*-2-hexene; It is a device measuring the angle (the amount of rotation) through which polarized light has been rotated when passing through a sample tube containing one or more optically active substances; They refer to the way that an optically active isomer rotates plane-polarized light, i.e., clockwise or to the right for dextrorotatory and counterclockwise or to the left for levorotatory.

07 7-bromo-4-methyl-5-octen-2-yn-1-ol.

08 (a) alcohol; (b) phenol; (c) ether; (d) carboxylic acid; (e) ester; (f) carboxylic acid anhydride.

09 (a) p-Ph(Cl)$_2$; (b) C—C—C=C—C—C—OH;
(c) C—C—C—C—C—C; (d) Br-Ph-CH$_3$;
(e) CH$_3$—CH$_2$—NH—CH$_2$—CH$_3$; (f) H$_3$CO—(o-)Ph—CH=O;
(g) C=C—C—H; (h) =(c);
(i) C—C(NH$_2$)—COOH; (j) C—C—C(CH$_3$)—CO—O—OC—C(CH$_3$)—C—C;
(k) m-Ph(NO$_2$)$_2$(COCl); (l) CH$_3$CO—N(CH$_3$)$_2$;
(m) Ph—Ph—CO—NH$_2$; (n) —(CH$_2$—CH$_2$(Ph))$_n$—;
(o) (CH$_2$—CH$_2$—O)$_n$—; (p) —(CH$_2$)$_n$—;
(q) (—NH(CH$_2$)$_5$CO)$_n$— and —[CO—(CH$_2$)$_4$—CO—NH(CH$_2$)$_6$NH]$_n$—.

10 (A—B)$_n$—; (A)$_n$—(B)$_m$1—;...—A—B—B—A—A—B—A—B—...

11 H$_2$N—CHR—COOH; Levorotatory optical isomer: H$_2$N—C*(H)(R)—COOH; In most living organisms, only L-amino acid can be catalyzed and synthesized by enzymes to form proteins—a process which may be largely due to natural selection.

Homework No. 8

01 (a) Carbohydrate is a general name for sugars, starches, and cellulose.

(b) Today chemists also refer to carbohydrates as saccharides, after the smaller units from which they are built.

(c) The most important monosaccharides are sugars glucose, fructose, and galactose, isomers with the general formula $C_6H_{12}O_6$. Each of these sugars can exist in either of two ring forms or in an open-chain form.

(d) Disaccharides are composed of two monosaccharide units and the examples are maltose, lactose and sucrose.

(e) Polysaccharides consist of many saccharide units linked together to form long chains. The most common polysaccharides are starch, glycogen (sometimes called animal starch), and cellulose. All of these are composed of repeating glucose units, but they differ in the way the glucose units are attached.

(f) All the polysaccharides are polymers, a general name for large molecules composed of repeating units called monomers.

(g) A monomer is the repeating unit of a polymer.

(h) Amino acids are linked together by a peptide bond, created when the carboxylic acid group pf one amino acid reacts with the amine group of another amino acid to form an amide functional group.

(i) The product of the reaction described in (h) is called a peptide. Although the language used to describe peptides is not consistent among scientists, small peptides are often called oligopeptides, and large peptides are called polypeptides.

(j) Because the reaction that links amino acids produces water as a by-product, it is an example of a condensation reaction, a chemical change in which a larger molecule is made from two smaller molecules accompanied by the release of water or another small molecule.

(k) All protein molecules are polypeptides.

(l) The primary structure of a protein is the linear sequence of its amino acids.

(m) The arrangement of atoms that are close to each other in the polypeptide chain is called the secondary structure of protein.

(n) When the long chains of amino acids link to form protein structures, not only do they arrange themselves into secondary structures, but the whole chain also arranges itself into a very specific overall shape called the tertiary structure of the protein. The protein chain is held in its tertiary structure by interactions between the side chains of its amino acids.

(o) Aanimal fats and vegetable oils are made up of triglycerides, which have many different structures but the same general design: long-chain hydrocarbon groups attached to a three-carbon backbone.

(p) A process called hydrogenation converts liquid triglycerides to solid triglycerides by adding hydrogen atoms to the double bonds and so converting them to single bonds.

(q) When enough hydrogen atoms are added to a triglyceride to convert all double bonds to single bonds, we call it a saturated triglycedride (or fat). It is saturated with hydrogen atoms.

(r) A triglyceride that still has one or more carbon-carbon double bonds is an unsaturated triglyceride.

02 (a) sugars, starches; (b) disaccharides, polysaccharides; (c) monosaccharide; (d) glucose; (e) galactose; (f) fructose; (g) glycogen, glucose; (h) energy, amylose, amylopectin; (i) liver, muscle cells; (j) repeating units; (k) cellulose; (l) monomers, amino acids; (m) peptide, amide; (n) water; (o) linear sequence; (p) close to each other; (q) shape; (r) long-term; (s) 37 kJ/g, 17 kJ/g; (t) hydrogenation; (u) single bonds; (v) double bonds; (w) partially.

03 (a) amino acid; (b) carbohydrate; (c) triglyceride; (d) steroid; (e) peptide.

04 (a) disaccharide; (b) monosaccharide; (c) monosaccharide; (d) polysaccharide.

05 (a) disaccharide; (b) monosaccharide; (c) polysaccharide; (d) polysaccharide.

06 Glucose and galactose differ in the relative positions of an —H and an —OH on one of their carbon atoms. In the standard notation for the open-chain form, glucose and galactose differ only in the relative position of the —H and —OH groups on the fourth carbon from the top. In the standard notation for the ring structures, the —OH group is down on the number 4 carbon of glucose and up on the number 4 carbon of galactose. See Figures 8.1 and 8.2.

07 Maltose—2 glucose units; Lactose—glucose and galactose; Sucrose—glucose and fructose.

08 Starch and cellulose molecules are composed of many glucose molecules linked together, but cellulose has different linkages between the molecules than starch. See Figure 8.4.

09 One end of the amino acid has a carboxylic acid group that tends to lose an H^+ ion, and the other end has a basic amino group that attracts H^+ ions. Therefore, in the conditions found in our bodies, amino acids are likely to be in the second form.

10

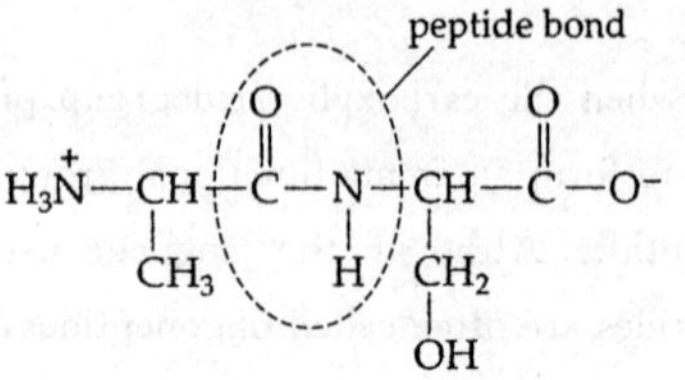

$$H_3\overset{+}{N}-CH-\underset{\underset{CH_3}{|}}{C}\overset{O}{\underset{\|}{-}}\overbrace{-N-CH}^{\text{peptide bond}}-\underset{\underset{\underset{OH}{|}}{CH_2}}{\overset{O}{\underset{\|}{C}}}-O^-$$

11

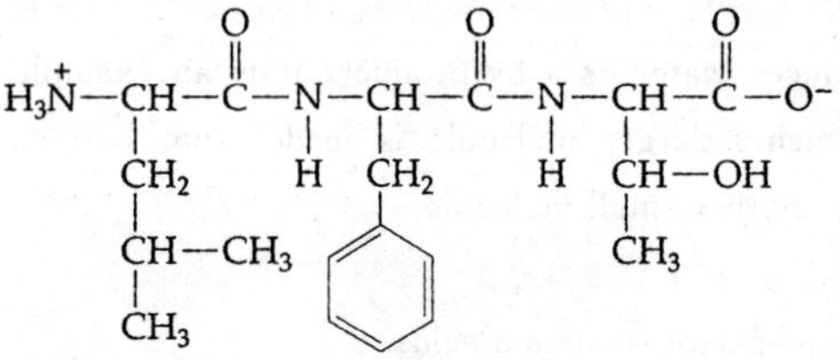

$$H_3\overset{+}{N}-\underset{\underset{\underset{CH_3}{|}}{\underset{CH-CH_3}{|}}}{CH}-\overset{O}{\underset{\|}{C}}-\underset{H}{N}-\underset{\underset{C_6H_5}{|}}{CH}-\overset{O}{\underset{\|}{C}}-\underset{H}{N}-\underset{\underset{\underset{CH_3}{|}}{CH-OH}}{CH}-\overset{O}{\underset{\|}{C}}-O^-$$

12

aspartic acid | phenylalanine | methanol

$$H_2N-\underset{\underset{\underset{\underset{OH}{|}}{C=O}}{CH_2}}{C}\underset{H}{|}\overset{O}{\underset{\|}{-C}}-N-\underset{\underset{C_6H_5}{|}}{C}\underset{H}{|}\overset{O}{\underset{\|}{-C}}-O-CH_3$$

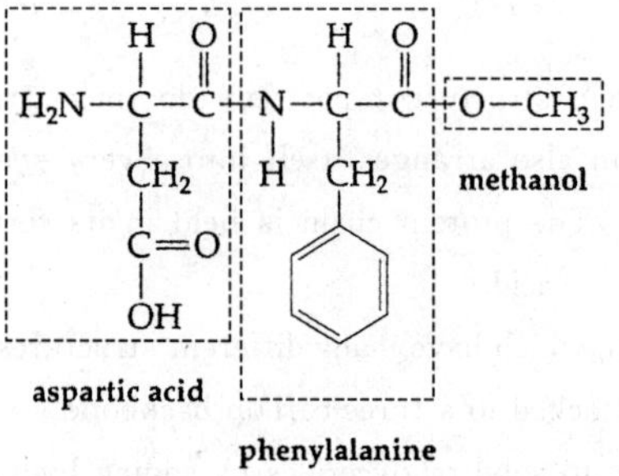

13 Each of these interactions draw specific amino acids in a protein chain close together, leading to a specific shape of the protein molecule. Disulfide bonds are covalent bonds between sulfur atoms from two cysteine amino acids (Figure 8.11). Hydrogen bonding forms between —OH groups in two amino acids, like serineor threonine, in a protein chain (Figure 8.12). Salt bridges are attractions between negatively charged side chains and positively charged side chains. For example, the carboxylic acid group of an aspartic acid side chain can lose its H^+, leaving the side chain with a negative charge. The basic side chain of a lysine amino acid can gain an H^+ and a positive charge. When these two charges form, the negatively charged aspartic acid is attracted to the positively charged lysine by a salt bridge (Figure 8.13).

14 The compound shown in the upper (lower) part of the figure is saturated (unsaturated) and more likely to be a solid (liquid) at room temperature.

15

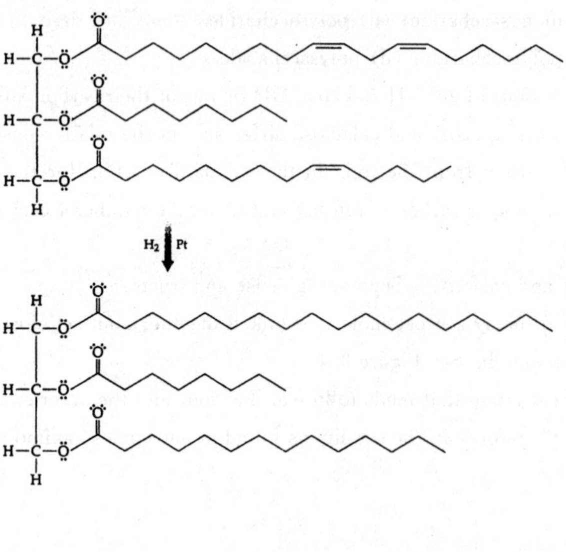

Homework No. 9

01 (a) When samll molecules, such as water, are released in the formation of a polymer, the polymer is called a condensation (or sometimes a step-growth) polymer.

(b) The polymerization reaction described in (a) is termed step-growth polymerization.

(c) Unlike condensation (step-growth) polymers, which release small molecules, such as waters, as they form, the reactions that lead to addition, or chain-growth, polymers incorporate all of the reactants' atoms into the final product.

02 (a) condensation; (b) step-growth; (c) parentheses, n; (d) diol; (e) addition.

03 W. H. Carothers, working for E. I. Du Pont de Nemours and Company, developed the first synthetic polyamide. He found a way to react adipic acid (a di-carboxylic acid) with hexamethylene diamine (which has two amine functnal groups) to form long-chain polyamide molecules called nylon 66 (the first "6" in the "66" indicates the number of carbon atoms in each portion of the polymer chain that are contributed by the diamine, and the second "6" shows the number of carbon atoms in each portion that are contributed by the di-carboxylic acid.) The reactants are linked together by condensation reactions in which an —OH group removed from a carboxylic functional group combines with a —H from an amine group to form water, and an amide linkage forms between the reacting molecules.

04 Polyethylene molecules can be made using different techniques. One process leads to branches that keep the molecules from fitting closely together. Other techniques have been developed to make polyethylene molecules with very few branches. These straight-chain molecules fit together more efficiently, yielding a high-density polyethylene, HDPE, that is more opaque, harder, and stronger than the low-density polyethylene, LDPE.

05 One of the reasons for the exceptional strength of nylon is the hydrogen bonding between amide functinal groups. A higher percentage of amide functional groups in nylon molecules' structures leads to stronger hydrogen bonds between them. Thus, changing the number of carbon atoms in the diamine and in the dicarboxylic acid changes the properties of nylon. Nylon 610, which has four more carbon atoms in the dicarboxylic acid molecules that form it than for nylon 66, is somewhat weaker than nylon 66 and has a lower melting point.

06 (a) polyethylene; (b) polyester; (c) polypropylene; (d) polystyrene; (e) nylon; (f) poly (vinyl chloride).

07 Nonpolar molecules are attracted to each other by London forces, and increased sizes of molecules lead to stronger London forces. Polyethylene molecules are much larger than the ethylene molecules that are used to make polyethylene, so polyethylene molecules have much stronger attractions between them, making them soilids at room temperature.

附录 C 试题举例
Appendix C Sample Mid-Term and Final Exams
C.1 Sample Mid-Term Exam

Physical and mathematical constants: (a) ideal gas constant $R = 0.0821$ L·atm/(K·mol) and 8.314 J/(K·mol); (b) 1 mol $= 6.022 \times 10^{23}$; (c) molar mass (g/mol) of H:1.0079; C:12.011; N:14.0067; O:15.9994; N:14; Cl:35.453; Zn:65.38; Ba:137.33; Zn:65.38; K:39.0983; Cr:51.996; Ca:40.08; Sr:87.62.

01 (5%) Answer the following questions (1% for each answer):

(1) What two properties of real gas molecules cause deviations from ideal gas behavior?

(2) State the first law of thermodynamics in equation form. Explain this equation in your own words.

(3) Given that the average speed of oxygen molecules at 25℃ is 4.44×10^2 m/s, is the average speed of nitrogen molecules at 25℃ higher or lower than that of oxygen molecules?

02 (2%) A container has a mass of 68.31 g empty and 93.34 g filled with water. Calculate the volume of the container using a density of 1.0000 g/cm³ for water. The container when filled with an unknown liquid had a mass of 88.42 g. Calculate the density of the unknown liquid.

03 (2%) The approximate radius of a hydrogen atom is 0.058 nm and of a proton is 1.5×10^{-12} m. Assuming both the hydrogen atom and the proton to be spherical, calculate the fraction of the space in an atom of hydrogen that is occupied by the nucleus. $V = (4/3)\pi r^3$ for a sphere.

04 (5%) What is the molarity (1 M = 1 mol/L) of a solution containing 100.00 g Ba(OH)$_2$·8H$_2$O dissolved in enough water to make exactly 1 L of solution?

05 (5%) Carbon tetrachloride, CCl$_4$, was formerly used as a dry cleaning fluid. What are the (a) molecular mass and (b) molar mass of this substance? (c) Calculate the mass of one molecule of CCl$_4$. How many (d) moles and (e) molecules of CCl$_4$ are present in 17.93 g of the compound? How many (f) carbon atoms and (g) chlorine atoms are present in 17.93 g of sample?

06 (4%) The van der Waals constants for carbon tetrachloride, CCl$_4$, are $a = 20.39$ L²·atm/mol² and $b = 0.1383$ L/mol. Find the pressure of a sample of if it one mole occupies 30.0 L at 77℃ (just slightly above the boiling point). Assume to obey the (a) ideal gas law and (b) van der Waals gas law $[(p + an^2/V^2)(V - nb) = nRT]$.

07 (5%) The radius of a typical molecule of gas is 0.2 nm. (a) Find the volume of a molecule assuming it to be spherical. $[V = (4/3)\pi r^3$ for a sphere.] (b) Calculate the volume actually occupied by a mole of these molecules. (c) If a mole of this gas is at STP, find the fraction of the volume of the gas actually occupied by the molecules. (d) Comment on your answer to (c) in view of the first statement summarizing the kinetic-molecular theory of an ideal gas.

08 (6%) A sheet of iron was galvanized (plated with zinc) on both sides to protect it from rust. The thickness of the zinc coating was determined by allowing hydrochloric acid to react with the zinc and collecting the resulting hydrogen. [Note: The acid solution contained an "inhibitor" (SbCl$_3$) which prevented the iron from reacting.]

$$Zn(s) + 2HCl(aq) \longrightarrow ZnCl_2(aq) + H_2(g)$$

Determine the thickness of the zinc plate from the following data: sample size = 1.50 cm × 2.00 cm; volume of dry hydrogen = 30.0 mL; temperature = 25℃; pressure = 747 Torr; and density of zinc = 7.11 g/cm³.

09 (6%) A standard qualitative analysis scheme for the separation and identification of Ba^{2+}, Sr^{2+}, and Ca^{2+} ions in solution is

$$[Ba^{2+}, Sr^{2+}, Ca^{2+}](NO_3)_2(aq) + K_2CrO_4(aq) \xrightarrow{CH_3COOH, NH_4(CH_3COO)}$$
$$BaCrO_4(s) + [Sr^{2+}, Ca^{2+}](NO_3)_2(aq) + 2KNO_3(aq) \text{ mixture}$$

$$[Sr^{2+}, Ca^{2+}](NO_3)_2(aq) + K_2CrO_4(aq) \xrightarrow{OH^-} SrCrO_4(s) + Ca(NO_3)_2(aq) + 2KNO_3(aq)$$

$$Ca(NO_3)_2(aq) + K_2CrO_4(aq) \longrightarrow CaCrO_4(s) + 2KNO_3(aq)$$

What is the composition of a mixture of these ions if 1.00 g of each of the precipitates are collected? Assume each precipitate to be completely insoluble.

10. (5%) When a welder uses an acetylene torch, it is the combustion of acetylene that liberates the intense heat for welding metal together. The equation for this process is

$$2C_2H_2(g) + 5O_2(g) \longrightarrow 4CO_2(g) + 2H_2O(g)$$

The heat of combustion of acetylene is -1300 kJ/mol. What amount of heat is liberated when 0.260 kg of C_2H_2 is burned?

11. (5%) Exactly 250 J of energy was removed from one gram samples of oxygen and ozone. The O_2 cooled from 502℃ to 229℃ and the O_3 cooled from 498℃ to 192℃. Which substance has the higher molar heat capacity?

12. (3%) Write the structural formula for the isomer of the saturated hydrocarbon having the molecular formula C_5H_{12} which shows a single peak in its proton NMR spectrum and name it by the IUPAC system.

13. (14%) Write the structural formula for each of the following (1% for each answer):
    (1) permanganate ion;
    (2) sulfite ion;
    (3) manganic ion;
    (4) dichromate ion;
    (5) dihydrogen phosphate ion;
    (6) silver sulfide;
    (7) cupric chloride;
    (8) sulfurous acid;
    (9) dinitrogen pentoxide;
    (10) 1,2-dichlorocyclohexane;
    (11) 1-buten-3-yne or vinylacetylene;
    (12) p-xylene;
    (13) ethylene glycol dimethyl ether or 1,2-dimethoxyethane;
    (14) benzoic anhydride.

14. (12%) Write the English name for each of the following:
    (1) CH_3CH_3; (2) $CH_2=CH-CH_3$; (3) $CH\equiv CCH=CH_2$; (4) $CH_2ClCH_2NH_2$; (5) C_4-ring (□);
    (6) $HOOCCH_2COOH$.

15. (14%) Write the definition for each of the following significant terms (1% for each answer):
    (1) periodic table;   (2) noble gas;   (3) resonance;
    (4) evaporation;   (5) osmotic pressure;   (6) transition state (activated complex);
    (7) instrumental analysis.

16. (7%) Give the significant term for each of the following definitions or descriptions (1% for each answer):
    (1) Which is equal to the square root of the quantity obtained by the division of the sum of the squares of absolute deviations by the number of times of measurements minus one.
    (2) Which is a process wherein the capacity of a substance to combine with a reagent is quantitatively measured.
    (3) Which is a chemically reactive atom or group of atoms that imparts characteristic properties to the family of organic compounds containing it.
    (4) Which contain carbon and hydrocarbons, including natural gas and petroleum.
    (5) Which are crosslinked polymers that are permanently rigid, they do not melt when heated.
    (6) The synthesis of large molecules from simpler components, which generally requires energy.
    (7) The sequence of amino acids in the polyamide chain.

C.2 Sample Final Exam

01 (15%) Translate into Chinese the following paragraphs reprinted from *Chemical Physics Letters* 269 (1997) 356～364.

(1) Title: Hinged and chiral polydiacetylene carbon crystals

(2) Author: Baughman R H et. al.

(3) Abstract: Structures and properties are calculated for hinged and chiral polydiacetylene carbon phases, which consist entirely of carbon chains found in organic polydiacetylenes. Diffraction and spectroscopic results are consistent with those for the controversial reported carbon phases of unknown structure, which might have accidentally formed by the reaction of acetylenic chains. Predicted properties include (1) a semiconductor-to-metal transition upon doping, (2) a soft deformation mode that results in negative linear and area compress-ibilities, negative Poisson ratios and expected shape-memory and ferroelastic behavior for the hinged phases, and (3) optical activity, second harmonic generation and torsional and transverse compressional piezoelectricity for the chiral phase.

(4) Introduction: Because of their novel observed properties, single-crystal organic polydiacetylenes have been the focus of concentrated research since their discovery by Wegner [1] nearly thirty years ago. The properties include per-chain mechanical properties approaching those of diamond [2], various types of reversible chromism (thermal, piezo-, photo- and solvato-) [3], unusually high third-order non-linear optical properties [4], and anisotropic photoconductivity [5], However, as is typical of linear polymers, their key properties are diluted by the effects of substitutes. This Letter describes ….

02 (15%) Translate into English the following paragraphs reprinted from *Polymer Bulletin (Beijing)* (4) (1995)234～236.

提要：本文简要介绍了几种经特殊加工制成的拉胀高聚物，粗线条地描述了目前国际上对拉胀聚合物的拉胀机理或拉胀性的研究，并展望了拉胀高分子材料的研究和应用前景。

拉胀聚合物(auxetic polymer)即具有负泊松比的聚合物的总称，是近几年才出现的新型高分子材料。它具有受拉膨胀(体积增加)和受压收缩(体积减小)的力学特性，即拉胀性(auxeticity)。本文就几种典型拉胀高聚物的制备及一般拉胀机理的研究作一简介，并对拉胀高分子材料的研究和应用前景予以展望。

总之，拉胀聚合物代表一个崭新的研究和应用领域。它是继高分子材料在接近达到原子键强度这一极限强度之后出现的又一线曙光。目前，世界各国的研究人员(包括作者本人领导的实验组)正通过分子水平上的化学合成来实现高强度拉胀高分子材料的开发和应用。可以相信这一天的到来为期不远了。

03 (65%) Read the attached article (*Chem. Commun* 2000, 1531～1532) carefully and then answer the following questions (50%) and write an abstract of the article (15%).

(01) (02%) What do the words "auxetic" and "auxetics" in the first paragraph mean?

(02) (02%) What shape does a synclastic doubly curved surface take?

(03) (02%) Who recently designed molecular auxetics ahead of the two authors?
    (A) R Lakes; (B) S C Rogers; (C) Y Tobe; (D) A Krebs; (E) Don't know.

(04) (08%) Which figures in Fig. 1 is a representation of part of a planar poly-phenylacetylene infinte network? And write down the structural formula of phenylacetylene and the repeat unit of polytriangles-3-yne.

(05) (04%) What is approximately the distance of the π-π interaction?

(06) (04%) Is polytriangles-1-yne auxetic? What limiting value of ν_{32} will polytriangles-∞-yne have?

(07) (04%) For polytriangles-7-yne, what is the reason behind its very highly negative Poisson ratios in the Ox_2-Ox_3 plane?

(08) (04%) What do the words "zeolite" and "conspicuous" in the six paragraph mean? Do some zeolites also show auxetic behavior?

(09) (06%) Write down part of the structural representation of graphyne.

(10) (06%) Write down a synthetic route for the synthesis of triangle-1-yne. Where can one find the information about the synthesis of polyphenylacetylene networks other than polytriangles-n-yne?

(11) (02%) According the authors of this paper, has any man-designed molecular auxetics been synthesized so far?

(12) (04%) Write down the general definition of ν_{ij} in terms of ε_i and ε_j.

(13) (02%) Which volume of *Chem. Commun.* does this paper appear? What does J. N. G. represent? Will you have to contact J. N. G. to discuss matters related to this paper?

(14) (15%) Write an abstract of the article.

04 (5%) Write down your comments on the course, including the textbook, the mid-term and final exams, the video shows, the in-class conversation show, the home-work, etc.

C.3 A Complete Sample Exam with Answers
Questions and Problems in Chemistry(50%)

Physical and mathematical constants: (a) ideal gas constant $R=0.0821$ L · atm/(K · mol) and 8.314 J/(K · mol); (b) 1 mol = 6.022×10^{23}; (c) molar mass (g/mol) of O: 16; N: 14; Al: 27; Cl: 35.5; C: 12; Mg: 24.3; Ag: 107.868.

01 (5%) Answer the following questions (1% for each answer):

(01) What two properties of real gas molecules cause deviations from ideal gas behavior?

(02) What is the special name given to the heat exchanged between a system and its surroundings under constant pressure conditions? What is the symbol used to represent this quantity?

(03) During a process, the volume of a system remained constant. What is the value of the work for the process? Will the heat exchanged between the system and its surroundings equal to ΔH or ΔE?

02 (2%) The average speed of oxygen molecules at 25°C is 4.44×10^2 m/s. What is the average speed of nitrogen molecules at this temperature?

03 (4%) An excess of $AgNO_3$ reacts with 100.0 mL of an $AlCl_3$ solution to give 0.275 g of AgCl. What is the molarity of the $AlCl_3$ solution?

$$AlCl_3(aq) + 3AgNO_3(aq) \longrightarrow 3AgCl(s) + Al(NO_3)_3(aq).$$

04 (4%) Fluorescein is produced by heating a mixture of phthalic anhydride and resorcinol at 190~200°C for 10 h in the presence of anhydrous $ZnCl_2$

$$C_6H_4(CO)_2O + 2C_6H_4(OH)_2 \longrightarrow C_{20}H_{12}O_5 + X.$$

(01) Identify the missing product, X, in the above equation and write the complete balanced equation.

(02) Show that the equation is an example of the law of conservation of mass.

05 (5%) A 0.241 g piece of magnesium metal was added to 100.0 g of dilute hydrochloric acid in a calorimeter. A temperature increase of 10.89°C was observed as a result of the reaction

$$Mg(s) + 2HCl(aq) \longrightarrow MgCl_2(aq) + H_2(g).$$

The specific heat of the $MgCl_2$-HCl solution is 4.21 J/(K · g) and the heat capacity of the calorimeter is 2.74 J/K. Calculate the heat of reaction per mole of Mg that reacts.

06 (2%) Write the structural formula for the isomer of the saturated hydrocarbon having the molecular

formula C_5H_{12} which shows a single peak in its proton NMR spectrum and name it by the IUPAC system.

07 (14%) Write the structural formula for each of the following (1% for each answer):

(01) permanganate ion;　　　　　　　　　　(02) sulfite ion;
(03) manganic ion;　　　　　　　　　　　　(04) dichromate ion;
(05) dihydrogen phosphate ion;　　　　　　　(06) silver sulfide;
(07) cupric chloride;　　　　　　　　　　　(08) sulfurous acid;
(09) dinitrogen pentoxide;　　　　　　　　　(10) 1,2-dichlorocyclohexane;
(11) 1-buten-3-yne or vinylacetylene;　　　　(12) p-xylene;
(13) ethylene glycol dimethyl ether or 1,2-dimethoxyethane;　(14) benzoic anhydride.

08 (14%) Write the definition for each of the following significant terms (1% for each answer):

(01) periodic table;　　　　　　　　　　　(02) noble gas;
(03) resonance;　　　　　　　　　　　　　(04) evaporation;
(05) osmotic pressure;　　　　　　　　　　(06) transition state (activated complex);
(07) instrumental analysis;　　　　　　　　(08) standard deviation;
(09) titration;　　　　　　　　　　　　　(10) functional group;
(11) fossil fuels;　　　　　　　　　　　　(12) thermoset polymers;
(13) anabolism;　　　　　　　　　　　　　(14) primary structure of a biopolymer.

Comprehension and Writing (50%)

01 (30%) Read the attached article carefully and then answer the following questions:

(01) (2%) What does the word "such" in the second sentence of the first paragraph mean?

(02) (4%) Which of the following is the correct description of the isoprene stated in the article?

(a) $CH_2(OH)-C(CH_3)=CH-CH_6PO_3$;　　(b) $CH_3-C(H_4PO_4)=CH-CH=CH-OH$;
(c) $CH_2=C(CH_4PO_3)-CH=CH-OH$;　　(d) $C(H_5PO_4)=C(CH_3)-C(OH)=CH-OH$.

(03) (2%) Where do the solutions of evolving chemicals reside?

(04) (4%) A terpenoid molecule is most likely

(a) an alkane;　(b) a phenol;　(c) an alkene;　(d) a ketone;　(e) an alkyne.

(05) (4%) What is meant by the word "replicase"? Is it a noun or adjective?

(06) (4%) Which of the following scientists is most likely able to give the correct definition of ribozyme?

(a) Bernal;　(b) Wächtershäuser;　(c) Bartel;　(d) Kauffman;　(e) Spiegelman.

(07) (3%) Who envisaged the "molecular evolution"?

(08) (3%) Give a synthetic route for 6-aminopurine from hydrogen cyanide.

(09) (4%) What do the words "guesswork" and "randomness" in the title refer to?

02 (18%) Read the attached article carefully and then write an abstract of the article.

03 (02%) Write your major comments on and suggestions for this exam and the whole course.

Guesswork and randomness (for now at least)[①]

What came before the RNA world? That the first living things would not have been recognizable as such is readily accepted. They may simply have been aggregations of organic molecules which, under the illumination of sunlight, allowed the evolution of particular species, RNA molecules perhaps.

---

① Reprinted from: *Nature*, vol. 372, p. 30, 3 Nov. 1994.

附录 C　试题举例(Sample Mid-Term and Final Exams)

Synthesis of the nucleotide adenine from HCN

Decades ago, Bernal J D suggested that such processes might be sustained on the surfaces of clay particles. A newer version of that picture, due to Günter Wächtershäuser, is that the surfaces would have been minerals such as pyrite, capable of binding and polymerizing simple organic molecules fashioned from five-carbon isoprene units carrying phosphate and hydroxyl substitutes[01].

One virtue of the model is that, with the conversion of polymerized isoprene units into glycols, simplified cell membranes could be formed along with accumulating chemicals. Wächtershäuser describes an intermediate stage of evolution in which membranes would enclose solutions of evolving chemicals between themselves and the solid substrate beneath.

More recently, the argument has been carried further by the identification of terpenoid molecules from petroleum and other geological sedimentary deposits with chemicals known to be constitutents of extant cell membranes[02].

So what was the actual course of events more than 3.5 billion years ago? Detailed reconstructions of the emergence of RNA, or alternatively of isoprene-based chemicals, becomes guesswork of a kind.

So why not guess from the outset? That is the basis of the now common interest in "molecular evolution", partly stimulated by an experiment of Spiegelman in the 1960s. The RNA gene for the replicase enzyme (a protein) of the $Q\beta$ bacteriophage was used repeatedly to replicate further copies of the viral RNA in several cycles, generating departures from the original sequence in the process. Selection for the replicase giving the most rapid replication of the whole virus eventually yielded a dramatically shorter enzyme.

Now that the production of random sequences of RNA is relatively simple (by producing random sequences of DNA in a synthesizer, and then converting them to RNA, for example) there have been demonstrations of how to evolve, with the help of a selection process, RNA molecules with particular properties—their binding to small molecules[03] or their enzymatic function as ribozymes[04].

The prophet of this approach is Stuart Kauffman, from the Sante Fe Institute, New Mexico. A patent originally in his name is now owned by the Darwin Institute in Seattle, Washington. Expectations among industrial molecular biologists are high. Whether the same approach will be applicable to real life is another matter.

[01]　Wächtershäuser G. *Microbiol. Rev.* 52, 452~458(1988)
[02]　Ourisson G & Nakata Y. *Chem. & Biol.* 1, 11~23(1994)
[03]　Ellington A D & Szostak J W. *Nature* 346, 818~822 (1990)
[04]　Bartel D P & Szostak J W. *Science* 261, 1411~1418(1993)

Answers to Part One of the Exam

01　(5% for 5 answers)

　　(01) forces of attraction between molecules & fraction of the total volume occupied by molecules. (As the pressure is increased, the volume of real gas tends to decrease more (less) as a result of the first (second) property.)

(02) enthalpy change; $\Delta H = Q_p$.

(03) $W = 0$; $Q_V = \Delta E$.

02 (2%) 4.75×10^2 m/s.

03 (4%) (c) 0.00640 mol/L.

04 (4%)

(01) $C_6H_4(CO)_2O + 2C_6H_4(OH)_2 \longrightarrow C_{20}H_{12}O_5 + 2H_2O$;

(02) $C_{20}H_{16}O_7 \longrightarrow C_{20}H_{16}O_7$.

05 (5%) -466 kJ/mol.

06 (2%) $C(CH_3)_4$; 2,2-dimethylpropane.

07 (14% for 14 answers)

(01) MnO_4^-; (02) SO_3^{2-}; (03) Mn^{3+};

(04) $Cr_2O_7^{2-}$; (05) $H_2PO_4^-$; (06) Ag_2S;

(07) $CuCl_2$; (08) H_2SO_3(aq); (09) N_2O_5;

(10) m-Cl_2Ph; (11) $CH_2{=}CH{-}C{\equiv}CH$; (12) p-$(CH_3)_2$-Ph;

(13) $CH_3OCH_2CH_2OCH_3$; (14) $Ph{-}(CO)O(CO){-}Ph$.

08 (14% for 14 answers)

(01) The periodic table groups the elements in order of increasing atomic number in such a way that elements with similar properties fall near each other. As the atomic number increases, the number of electrons in each atom also increases.

(02) A noble gas is an element in which all energy sublevels that are occupied are completely filled.

(03) Resonance refers to the arrangement of valence electrons in molecules or ions for which several Lewis structures can be written.

(04) Evaporation is the escape of molecules from a liquid in an open container to the gas phase.

(05) Osmotic pressure is defined as the external pressure exactly sufficient to oppose osmosis and stop it.

(06) The short-lived combination of reacting atoms, molecules, or ions that is intermediate between reactants and products is called the transition state or activated complex.

(07) Instrumental analysis relies upon optical, electrochemical, and other physical or physicochemical properties of sample solutions.

(08) The standard deviation is equal to the square root of the quantity obtained by the division of the sum of the squares of absolute deviations by the number of times of measurements minus one.

(09) A titration is a process wherein the capacity of a substance to combine with a reagent is quantitatively measured.

(10) A functional group is a chemically reactive atom or group of atoms that imparts characteristic properties to the family of organic compounds containing it.

(11) Fossil fuels contain carbon and hydrocarbons, including natural gas and petroleum.

(12) Thermoset polymers are crosslinked polymers that are permanently rigid, they do not melt when heated.

(13) The synthesis of large molecules from simpler components, which generally requires energy, is called anabolism.

(14) Primary structure of a biopolymer refers to the sequence of amino acids in the polyamide chain. (Details of the microstructure of a chain is a description of the primary structure.)

Answers to Part Two of the Exam

01 (30%)

(01) (2%) RNA;

(02) (4%) (D);

(03) (2%) between membranes and the substrate;

(04) (4%) (C);

(05) (4%) a noun;

(06) (4%) (C);

(07) (3%) Stuart Kauffman;

(08) (3%) the same as the synthesis shown in the article;

(09) (4%) The word "guesswork" here refers to detailed reconstructions of the emergence of RNA, or alternatively of isoprene-based chemicals, while "randomness" the evolution of RNA molecules with particular properties. As a whole, they refer to the state of affairs in the searching of a answer to the question "What came before the RNA world".

02 (18%) Several models are briefly discussed which try to explain the origin of life on the earth. One involves detailed reconstructions of the emergence of RNA or isoprene-based chemicals while the other the evolution of RNA molecules with particular properties by means of a selection process.

03 (2%) This exam is not an easy (or difficult) one. I like (or dislike) the course very much. To further improve the exam and the course, I suggest that...(omitted)

附录 D 科技阅读和翻译课文
Appendix D Useful Texts for Scientific Reading and Translations

D.1 Selected Speed Reading Texts

Reading (01) The Scientific Attitude

READING PASSAGE

What is the nature of the scientific attitude, the attitude of the man or woman who studies and applies physics, biology, chemistry, geology, engineering, medicine or any other science?

We all know that science plays an important role in the societies in which we live. Many people believe, however, that our progress depends on two different aspects of science. The first of these is the application of the machines, products and systems of applied knowledge that scientists and technologists develop. Through technology, science improves the structure of society and helps man to gain increasing control over his environment. New fibres and drugs, faster and safer means of transport, new systems of applied knowledge (psychiatry, operational research, etc.) are some examples of this aspect of science.

The second aspect is the application by all members of society, from the government official to the ordinary citizen, of the special methods of thought and action that scientists use in their work.

What are these special methods of thinking and acting? First of all, it seems that a successful scientist is full of curiosity—he wants to find out how and why the universe works. He usually directs his attention towards problems which he notices have no satisfactory explanation, and his curiosity makes him look for underlying relationships even if the data available seem to be unconnected. Moreover, he thinks he can improve the existing conditions, whether of pure or applied knowledge, and enjoys trying to solve the problems which this involves.

He is a good observer, accurate, patient and objective and applies persistent and logical thought to the observations he makes. He utilizes the facts he observes to the fullest extent. For example, trained observers obtain a very large amount of information about a star (e.g., distance, mass, velocity, size, etc.) mainly from the accurate analysis of the simple lines that appear in a spectrum.

He is sceptical—he does not accept statements which are not based on the most complete evidence available—and therefore rejects authority as the sole basis for truth. Scientists always check statements and make experiments carefully and objectively to verify them.

Furthermore, he is not only critical of the work of others, but also of his own, since he knows that man is the least reliable of scientific instruments and that a number of factors tend to disturb impartial and objective investigation.

Lastly, he is highly imaginative since he often has to look for relationships in data which are not only complex but also frequently incomplete. Furthermore, he needs imagination if he wants to make hypotheses of how processes work and how events take place.

These seem to be some of the ways in which a successful scientist or technologist thinks and acts.

COMPREHENSION

01. Name some sciences.
02. Name two ways in which science can help society to develop.
03. Give some examples of the ways in which science influences everyday life.
04. What elements of science can the ordinary citizen use in order to help his society to develop?
05. How can you describe a person who wants to find out how and why the universe works?
06. What is the role of curiosity in the work of a scientist?
07. Name some of the qualities of a good observer.
08. Give an example of how observed facts are utlized to the fullest.
09. How does a sceptical person act?
10. How does the scientist act towards (a) evidence presented by other people, (b) evidence which he presents in his own work?
11. What do you know about the data which the scientist often has to use? How does this affect his way of thinking?
12. For what other purposes does a scientist need imagination?

DISCUSSION AND CRITICISM

01. Do you think there are other special ways of thinking and acting, used by scientists?
    If so, comment and explain.
02. Do you think some of these ways are more important than others? If so, give reasons.
03. Do you know of any famous scientist whose work demonstrates some or all the qualities mentioned in the passage. Give details.
04. In what ways do other sciences affect the particular science you study yourself?
    Give examples.
05. Do you agree that it is important to train the non-scientist to think in a scientific way.
    Give good evidence for your point of view.
06. Do you agree that "man is the least reliable of scientific instruments"? Give examples.
07. Give a clear explanation of what you think the word "authority" means.

Reading (02)   Scientific Method and the Methods of Science

READING PASSAGE

It is sometimes said that there is no such thing as the so-called "scientific method"; there are only the methods used in science. Nevertheless, it seems clear that there is often a special sequence of procedures which is involved in the establishment of the working principles of science. This sequence is as follows: (1) a problem is recognized, and as much information as appears to be relevant is collected; (2) a solution (i.e., a hypothesis) is proposed and the consequences arising out of this solution are deduced; (3) these deductions are tested by experiment, and as a result the hypothesis is accepted, modified or discarded.

As an illustration of this we can consider the discovery of air-pressure. Over two thousand years ago, men discovered a method of raising water from one level to another by means of the vacuum pump. When, however, this machine passed into general use in the fifteenth and sixteenth centuries, it was discovered that, no matter how perfect the pump was, it was not possible to raise water vertically more than about 35 feet. Why? Galileo, amongst others, recognized the problem, but failed to solve it.

The problem was then attacked by Torricelli. Analogizing from the recently-discovered phenomenon of water-pressure (hydrostatic pressure), he postulated that a deep "sea of air" surrounded the earth; it was, he thought, the pressure of this sea of air which pushed on the surface of the water and caused it to rise in the vaccum tube of a pump. A hypothesis, then, was formed. The next step was to deduce the consequences of the hypothesis. Torricelli reasoned that this "air pressure" would be unable to push a liquid heavier than water as high as 35 feet, and that a column of mercury, for example, which weighed about 14 times more than water, would rise to only a fourteenth of the height of water, i. e. approximately 2.5 feet. He then tested this deduction by means of the experiment we all know, and found that the mercury column measured the height predicted. The experiment therefore supported the hypothesis. A further inference was drawn by Pascal, who reasoned that if this "sea of air" existed, its pressure at the bottom (i. e., sea-level) would be greater than its pressure further up, and that therefore the height of the mercury column would decrease in proportion to the height above sea-level. He then carried the mercury tube to the top of a mountain and observed that the column fell steadily as the height increased, while another mercury column at the bottom of the mountain remained steady (an example of another of the methods of science, the controlled experiment). This further proof not only established Torricelli's hypothesis more securely, but also demonstrated that, in some aspects, air behaved like water; this, of course, stimulated further enquiry.

COMPREHENSION

01. What does the establishment of the working laws of science often involve?
02. What does a scientist collect when he tries to establish a scientific law?
03. What is the next step in the process described above?
04. What does the scientist then deduce?
05. How does he proceed to verify these deductions?
06. What does he finally do with his original hypothesis?
07. Give an approximate date for the invention of the vacuum pump.
08. Is it possible to raise water from the bottom floor of a building to the roof 50 feet above, using a vacuum pump? Why?
09. What was Torricelli's theory about the height of the water in a vacuum tube?
10. What were his deductions, concerning the effect of air pressure on a column of mercury?
11. What further inference was made by Pascal?
12. Why did he use two mercury tubes?
13. What were the three results of Pascal's experiment?

14. What do you think happened to the mercury column when it was carried down the mountain?

DISCUSSION AND CRITICISM

01. Give an example of scientific method used in the development of the science you study yourself.
02. Do you agree that there is no one scientific method? Give reasons and examples.
03. What do you think is meant by "as much information as appears to be relevant is collected"? What was the relevant information in Torricelli's case? (Note the developments in hydrostatics.)

Reading (03) Sources of Error in Scientific Investigation

READING PASSAGE

In Reading (02) we examined briefly the sequence of procedures which make up the so-called scientific method. We are now going to consider a few of the many ways in which a scientist may fall into error while following these procedures.

In formulating hypotheses, for example, a common error is the uncritical acceptance of apparently common-sense, but untested, assumptions. Thus in the field of psychology it was for many years automatically assumed that the main cause of forgetfulness was the interval of time elapsing between successive exposures to a learning stimulus. Experimentation, however, was subsequently undertaken, and several other factors, such as motivation and strength or effectiveness of the stimulus, turned out to have an even more important bearing on the problem. A somewhat similar error arises from neglect of multiple causes. Thus two events may be found to be associated, e.g., when the incidence of a disease in a smoky industrial sector of a city is significantly higher than in the smoke-free zones. A research worker might infer that the existence of the disease is due to the smokiness of the area when in fact it might equally well be found in other reasons, such as the under-nourishment of the inhabitants or over-crowding.

Both in collecting the original evidence and in carrying out subsequent experiments, a frequent cause of error is the fact that observations are not continued for a long enough time. This may lead not only to a failure to discover positive items (e.g., Le Monnier's failure to recognize that Uranus was a new planet, not a fixed star, etc.), but may also result in important negative aspects of the investigation remaining undiscovered. In applied science, this latter error may have disastrous consequences, as in the case of the thalidomide drugs, cancer-inducing industrial chemicals, etc.

Another well-known error in experimentation is lack of adequate controls. Thus a few years ago it was widely believed that a certain vaccine could prevent the common cold, since in the experiments the vaccinated subjects reported a decrease in the incidence of colds compared with the previous year. Yet later, more strictly-controlled experiments failed to support this conclusion, which could have been due to a misinterpretation of chance results. This error is often caused by a failure to test a sufficient number of subjects (inadequate sampling), a disadvantage which affects medical and psychological research in particular.

Errors in measurement, particularly where complicated instruments are used, are common: they may arise through lack of skill in the operator or may be introduced through defects in the apparatus itself. Furthermore, it should be borne in mind that apparently minor changes in laboratory conditions, such as variations in the electric current, or failure to maintain atmospheric conditions constant, may disturb the accuracy of various items of equipment and hence have an adverse influence on the experiment or series of experiments as a whole. In addition, such errors tend to be cumulative.

Finally, emotion in the observer can be one of the most dangerous sources of error. This may cause the researcher to over-stress or attach too much importance to irrelevant details because of their usefulness in supporting a theory to which he is personally inclined. Conversely, evidence disproving the view held may be ignored for similar reasons. Even routine matters such as the recording of data may be subject to emotional interference, and should be carefully checked.

To sum up (summarize), the multiple possibilities of error are present at every stage of a scientific investigation, and constant vigilance (care) and the greatest foresight must be exercised in order to minimize or eliminate them. Additional errors are, of course, connected with faulty reasoning; but so widespread and serious are the consequences that may arise from this source that they deserve separate treatment in the following passage.

COMPREHENSION

01. What is the connection between the reading passage here and that of Reading (02)?
02. At what stage of an investigation is the scientist most likely to commit the error of accepting untested assumptions?
03. Give an example of this type of error.
04. What type of error is similar to the above?
05. Name at least three factors that an unduly high incidence of disease in a smoky sector of a city might be due to.
06. Name two broad results that insufficient observation may lead to. Give examples of each.
07. Why was it believed that a certain vaccine could cure the common cold?
08. What is meant by inadequate sampling?
09. Name two causes of inaccurate measurements.
10. What other factors can affect the accuracy of instruments?
11. What is an additional danger in these so-called minor errors?
12. Name three ways in which emotion can cause scientists to make mistakes.
13. How can the possibilities of error be minimized?
14. What is the last source of error named in the passage?

DISCUSSION AND CRITICISM

01. Can you suggest why medical and psychological researchers (rather than, say, entomologists) are liable to fall into the error of inadequate sampling?
02. Explain clearly what is meant by "lack of adequate controls".
03. Explain clearly how insufficient observation has led to a failure to reveal weaknesses in a theory or piece of applied research. As examples of the latter, you could consider (a) the thalidomide drugs; (b) cancer-inducing industrial chemicals; (c) modern pesticides and their often disastrous side-effects; (d) aircraft or other engineering failures; (e) any other examples you know or can find out for yourself.

04. Explain why the last line of the exercise in the Structure Study section recommends the student to try and find out the true stories of pieces of successful research. Why may these differ from the published accounts of the investigations?
05. Can you think of any further sources of error not mentioned in the Reading Passage and the exercise in the Structure Study section?

Reading (04) The Role of Chance in Scientific Discovery

READING PASSAGE

Nearly a century and a half ago, a Danish physicist, Oersted, was demonstrating current electricity to a class, using a copper wire which was joined to a Voltaic cell. Amongst the miscellaneous apparatus on his demonstration bench there happened to be a magnetic needle, and Oersted noticed that when the hand holding the wire moved near the needle, the latter was occasionally deflected. He immediately investigated the phenomenon systematically and found that the strongest deflection (deviation) occured when he held the wire horizontally and parallel to the needle. With a quick jump of imagination (or intuition, as it is often called when it produces successful results) he then disconnected the ends of the wire and reconnected them to the opposite poles of the cell—thus reversing the current—and found that the needle was deflected in the opposite direction. This chance discovery of the relationship between electricity and magnetism not only led quickly to the invention of the electric dynamo and hence to the large-scale utilization of electric energy, but forms the basis for modern electro-magnetic field theory, which is now an extremely valuable tool in both macro- and micro-physics.

The above story illustrates the part played in scientific discovery by chance (accident). Again, about 20 years ago a group of British bacteriologists and biochemists working in agricultural research were carrying out investigations into substances of organic origin which could be used to stimulate plant growth. One of the approaches they used consisted in studying the nodules (small round lumps) found on the root-hairs of certain plants, and which contain colonies of nitrogen-forming bacteria. Working on the hypothesis that these bacteria manufactured a substance which stimulated the nodule-forming tissue, the investigators eventually succeeded in isolating this substance. However, when they then tested it on various other plants, they found—quite contrary (opposite) to their expectations—that it actually prevented (inhibited) growth. Further systematic investigation showed that this toxic (poisonous) effect was selective, being much greater against dicotyledon (usually dicots and monocots respectively in USA) plants, which happen to include the majority of weeds, than against the monocotyledons, which include the grain crops and grasses. The researchers thus realized that they had discovered a powerful selective weed-killer: they continued their research, using inorganic compounds of related chemical composition, and in this way laid the foundations of a technology which is of the greatest value in present-day agriculture.

Another well-known instance of the role of chance is connected with the discovery of penicillin by Fleming. This medical researcher had been investigating some pathogenic (disease-causing) bacteria, and after being absent from his laboratory for some days found on his return that one of the culture dishes in which colonies of the bacteria were growing had been contaminated by a colony of another organism, a mould of penicillium SPP. He was going to throw the dish away when he noticed that the penicillium colony was surrounded by an area completely clear of the pathogenic bacteria. He immediately realized that the penicillium must have manufactured a substance which had broken down (disintegrated) the pathogenes. He then isolated this substance, which turned out to be the most powerful agent yet discovered against bacteria causing a number of dangerous and widely-spread diseases.

Apart from demonstrating the way in which chance may lead to scientific discoveries of primary importance, an analysis of the three cases outlined above may be useful in showing how a successful worker utilizes these accidental opportunities. The first point to notice is that although in all cases the key phenomenon produced results which were both unexpected and—in the last two cases—even apparently disadvantageous, the scientists invariably reacted in an extremely positive manner. The refusal to be disturbed or disorganized by unexpected or apparently adverse occurrences, but, on the contrary, to be stimulated by them, has in fact been a marked (strong) characteristic of successful investigators.

Secondly, we note that in the first and third cases the phenomena were very slight and might easily have escaped notice, whilst in the second case they produced a negative result. From this we might deduce that a superior capacity for observation is also a property of outstanding researchers. On this point, however, a psychologist would probably tend to disagree. He would point out that observation or perception is a concept which refers not so much to acuteness of sight, hearing, etc., or to the care with which they are applied, as to the ability to relate phenomena to a complex network of previous experiences and theories, i.e., to a meaningful frame of reference. In other words, an observer who lacks such a frame of reference will be unable to realize the significance of certain phenomena even though his senses may "experience" them, and so he may fail to observe them. This can be illustrated by the following example:

At the end of last century, an American chemist, Hillebrand, was using a recently-developed instrument, the spectroscope, to analyse the gas given off by a certain mineral when treated with acid. This instrument works on the principle that each individual substance emits a characteristic spectrum of light when its molecules are caused to vibrate by the application of heat, electricity, etc.; and after studying the spectrum which he had obtained on this occasion, Hillebrand reported the gas to be nitrogen. At this same time, another scientist, Rayleigh, happened to be investigating the anomalous fact that nitrogen obtained from the air appeared to be heavier than that obtained from other sources, e.g., ammonia (NH_3). Rayleigh repeated Hillebrand's experiment and, immediately noticing that the spectrum showed several bright lines which were additional to those typical of nitrogen, went on to discover the rare gases argon (Ar) and helium (He). Why had Rayleigh observed these extra lines whereas (while) Hillebrand apparently had not? Part of the answer seems to be that the former already possessed a frame of reference which included the possibility that a different sort of N might exist; he was therefore extremely sensitive to any apparent anomaly in the behaviour of this element. Hillebrand lacked this concept, and was therefore unable to see the slight deviant reaction of the gas he assumed he was dealing with.

This dual (double) quality of being sensitive to, and curious about, small accidental occurrences, and of possessing a frame of reference capable of suggesting their true significance, is probably what Pasteur meant when he said "Chance benefits only the prepared mind". Nevertheless, it is clear (plain, obvious) that these qualities alone, even when joined to those mentioned previously, are not necessarily sufficient to ensure success: an indispensable factor in all the discoveries quoted above was careful and systematic experimentation. We may therefore conclude that it is the capacity to plan and undertake such experimentation which finally allows the investigator to make the most of his luck—if it comes.

COMPREHENSION

01. What was the accidental phenomenon which Oersted noticed and investigated?
02. How did he make the needle deviate to the opposite direction to that of its original deflection?
03. What forms the basis of modern field theory?
04. What substance did the British agricultural researchers succeed in isolating?
05. Why were its effects on dicots of great interest to the investigators?
06. What are pathogens?
07. What evidence did Fleming find which led him to assume that the penicillium broke down the pathogens?
08. Besides illustrating the role of accident in scientific investigation, what else can we learn from the cases quoted?
09. Describe a strong characteristic of successful researchers which is demonstrated in each of the examples given.
10. What deduction regarding researchers might a psychologist disagree with?
11. Describe the concept of observing.
12. What is a spectroscope?
13. Why was Rayleigh interested in nitrogen?
14. What phenomena led him to discover the inert gases argon and helium?
15. What did Pasteur probably mean by the prepared mind?
16. What additional capacity is usually necessary for the successful exploitation of accidental occurrences?
17. Give words meaning approximately the same as: to disintegrate; opposite; chance; marked; imagination; to inhibit; plain; deviation; whereas.

DISCUSSION AND CRITICISM

01. Describe in an orderly and accurate way an instrument or piece of apparatus which is used in the science you are studying. Then prepare clear and detailed instructions for its use, employing diagrams where necessary.
02. Do you agree with the remarks about *observation* given in the passage? If so, how do you think a scientist can acquire the "wide frames of reference" required?

03. Analysing the cases outlined in the Reading Passage, do you think they illustrate other qualities necessary for successful investigation, apart from those mentioned in the text?

04. Describe any examples known to you which illustrate Pasteur's saying that "Chance favours only the prepared mind".

D.2 Texts for English-to-Chinese Translation

Translating (01) The Origins of Chemistry

Chemistry is a very old science. The first person who saw a tree burn after it was struck by lightning was observing a chemical change. People learned many centuries ago to cook food and to make pottery vessels by strongly heating clay which they had molded into appropriate shapes.

Modern chemistry grew out of ancient technology, such as the smelting of ores, as well as out of medicine and alchemy. In addition, observations of the world around them led those who were philosophically minded to speculate about the nature of matter. Unfortunately, much of this speculation was far from the mark and actually retarded the progress of science. For example, some philosophers of the sixth century before Christ believed that water was the basis of all substances; others thought air was the primordial substance.

Plato (427~347 B.C.) postulated that there were four elements—earth, air, fire, and water, and perhaps a fifth—an ether which was in some way associated with the material of the heavens. Aristotle (384~322 B.C.) also thought that all matter was composed of these elements, a belief that persisted in science until the end of the eighteenth century.

Alchemy grew out of both technology and philosophical speculation. At first alchemy was concerned with the conversion of base metals into gold, and several experiments were cited to support the belief that this was possible. For example, when impure gold is strongly heated, especially in a molten salt, the impurities are burned away, leaving pure gold. The resulting change in color and other properties was interpreted to mean that the original metal had changed into gold.

Physicians of early times tested all sorts of materials for their medicinal value, not on mice and guinea pigs, as is done today, but on their human patients. Herbs and vegetable extracts were used, as well as inorganic salts and minerals. Some valuable drugs were discovered in this way, and also some deadly poisons! Many physicians practiced alchemy on the side, and eventually alchemy came to include a search for the "elixir of life", a substance that would ensure perpetual youth and health. There were many opportunities for "get rich quick" schemes, and some alchemists took advantage of these. On that account, we often think of the alchemists as quacks and charlatans. Doubtless many of them were, but others were earnest seekers after truth.

The Greek philosophers drew some sound conclusions about the nature of matter, and many alchemists correctly recorded the physical and chemical properties of many substances.

However, their scientific progress was limited. The philosophers did few experiments, and although the alchemists did many, none involved accurate measurements.

Most areas of science are not considered fully explored until they become quantitative, that is, until accurate measurements are made. To learn about the composition of matter and about changes in the composition of matter, experiments must be performed, most often on weighed amounts of material under carefully controlled conditions. After the changes are complete, the products formed must be weighed before conclusions can be drawn about what has happened. By the middle of the eighteenth century the need for quantitative experiments had been recognized, and progress was aided by the availability of finely made, accurate balances constructed to meet the need. A study of the chemistry of carbon dioxide done by Joseph Black in 1754 has been called "the first example we possess of a clear, reasoned series of chemical researches where nothing was taken on trust, but everything was made the subject of careful, quantitative measurement".

Translating (02)　How to Solve It

It would be a mistake to think that solving problems is a purely "intellectual affair"; determination and emotions play an important role. Lukewarm determination and sleepy consent to do a little something may be enough for a routine problem in the classroom. But, to solve a serious scientific problem, will power is needed that can outlast years of toil and bitter disappointments.

Determination fluctuates with hope and hopelessness, with satisfaction and disappointment. It is easy to keep on going when we think that the solution is just around the corner; but it is hard to persevere when we do not see any way out of the difficulty. We are elated when our forecast comes true. We are depressed when the way we have followed with some confidence is suddenly blocked, and our determination wavers.

"Il n'est point besoin espérer pour entreprendre ni réussir pour persévérer." "You can undertake without hope and persevere without success." Thus may speak an inflexible will, or honor and duty, or a nobleman with a noble cause. This sort of determination, however, would not do for the scientist, who should have some hope to start with, and some success to go on. In scientific work, it is necessary to apportion wisely determination to outlook. You do not take up a problem, unless it has some interest; you settle down to work seriously if the problem seems instructive; you throw in your whole personality if there is a great promise. If your purpose is set, you stick to it, but you do not make it unnecessarily difficult for yourself. You do not despise little successes, on the contrary, you seek them; if you cannot solve the proposed problem try to solve first some related problem. …

Incomplete understanding of the problem, owing to lack of concentration, is perhaps the most widespread deficiency in solving problems. With respect to devising a plan and obtaining a general idea of the solution two opposite faults are frequent. Some students rush into calculations and constructions without any plan or general idea; others wait clumsily for some idea to come and cannot do anything that would accelerate its coming. In carrying out the plan, the most frequent fault is carelessness, lack of patience in checking each step. Failure to check the result at all is very frequent; the student is glad to get an answer, throws down his pencil, and is not shocked by the most unlikely results.

D.3　Texts for Chinese-to-English Translation

Translating (01)　拉胀性高分子材料[①][②]
(Auxetic Polymer Materials)

提要　本文简要介绍了几种经特殊加工制成的拉胀高聚物,粗线条地描述了目前国际上对拉胀聚合物的拉胀机理或拉胀性的研究,并展望了拉胀高分子材料的研究和应用前景。

关键词　拉胀聚合物,拉胀性,拉胀物

拉胀聚合物(auxetic polymer)即具有负泊松比的聚合物的总称,是近几年才出现的新型高分子材料。它具有受拉膨胀(体积增加)和受压收缩(体积减小)的力学特性,即拉胀性(auxeticity)。本文就几种典型拉胀高聚物的制备及一般拉胀机理的研究作一简介,并对拉胀高分子材料的研究和应用前景予以展望。

1. 拉胀高聚物举例

虽然关于负泊松比固体的报道可上溯至19世纪下半叶[01],但作为具有实际用途的材料问世,还应首推泊松比在-0.7左右的聚氨酯泡沫材料[02,03]。这是一种对传统高分子泡沫材料的结构作了某种改变后所得到的新型高分子泡沫材料,且不局限于聚氨酯类热塑性高分子。可用来制备这类拉胀泡沫材料的还有热固性硅氧烷弹性体和金属铜[04]。这类拉胀高聚物的主要结构特征表现在所谓的凹式微结构(reentrant microstructure)上,即内含通过将一般多面体若干个顶点下凹后得到的伞式星状微结构。其性能除了由负泊松比这一材料性质所带来的不同之外,还具有比一般高分子泡沫材料高得多的弹性。

另一类拉胀高聚物系由对现有高分子品种进行特殊加工处理后制得,包括各向异性的多孔聚四氟乙烯[05]和超高分子量聚乙烯[06,07]。前者的制备涉及对烧结聚四氟乙烯颗粒进行快速加热和拉伸[08],所得制品主要用作半渗透膜,还可应用于生物医学工程。英国学者Caddock和Evans[05]首次对该高聚物进行泊松比测试,发现该材料具有负的泊松比,其值随应变而变化,可达-12,即某一方向上单位伸长可引起其垂直方向上约12倍的伸展(注:各向异性材料的泊松比可小于-1)。拉胀聚乙烯的制备要求较高,主要包括压型、烧结和挤出三个步骤,所得制品也系多孔材料,但却是均质和连续的,且有较高的模量。通过对其圆柱制品进行沿径向方向的压缩试验,发现其泊松比在-1.24~0之间。

最后要介绍一种高分子水凝胶,它在接近发生体积相变时呈现负的泊松比,其值随温度而变化,这便是聚N-异丙基丙烯酰胺的水凝胶[09]。它是由单体N-异丙基丙烯酰胺和交联剂N,N'-亚甲基双丙烯酰胺在水溶液中进行自由基聚合而制得。该凝胶的泊松比ν根据实验测得

---

①　选自:《高分子通报》(1995)、《当代化学前沿》(1998)、《中华新论·全国优秀科技理论研究成果信息库》(2000)和《21世纪中国改革发展论坛文集》(2001);作者:魏高原。

②　因篇幅所限,本文部分参考文献[04~33]略去。

的本体模量 K 与剪切模量 μ 算出,即 $\nu=(3K-2\mu)/(6K+2\mu)$。显然,在临界点($K=0$)上,泊松比应为-1;而当凝胶的 K 接近于零而 μ 保持几乎不变时,ν 为负值,即凝胶表现出拉胀性。

2. 拉胀机理

关于拉胀聚合物的拉胀机理的研究到目前为止主要限于对其微结构的拉胀行为的探索,而这种探索又有对微结构的规整排列与无规连接之分。前者主要见于 Evans 等人[10~21]的工作,后者则由本文作者等人[22~24]完成。此外,关于平面各向同性网络(柔性或刚性)或团粒(圆形或正六边形)材料以及内含拉胀性包埋物的聚合物基复合材料的负泊松比的产生机理也有诸多报道[25~38]。

尽管不同类型的拉胀高聚物其拉胀机理各有不同,但有一点却是共同的,即材料宏观上所表现出的拉胀行为是由其微观上的种种机制采用某种确定的合作方式来完成的。例如,伞撑开(沿支柱方向拉伸)后体积增大而收拢后体积减小便可看作是由于各支架沿支柱的滑动这一协同效应而产生的拉胀现象。目前,对高聚物拉胀机理的研究正方兴未艾。

3. 应用前景

拉胀高分子材料的用途由其所具有的负泊松比这一材料性质所决定。对于三维各向同性材料,负的泊松比意味着高剪切模量[$\mu=E/(2+2\nu)$,E 为杨氏模量],这无疑可改善柱状和层状结构的抗风抗震性能[33~35]。拉胀高聚物用做铺路材料,可显著增强耐压抗震性能。在海洋深水作业方面,拉胀高分子材料将表现出很高的液压稳定性。制造太空飞行器表层所用的纤维增强复合材料将会由于拉胀纤维的替代而使其抗裂强度显著提高[34,35]。此外,用拉胀高聚物制成的衣物、睡袋等日常用品会有良好的宽松舒适感,而将拉胀高分子用于制造人工血管时,其管壁将因血液流过时产生的切应力作用而变厚,耐用性因而大大提高[34]。

总之,拉胀聚合物代表一个崭新的研究和应用领域。它是继高分子材料在接近达到原子键强度这一极限强度之后出现的又一线曙光。目前,世界各国的研究人员包括作者本人领导的实验组正通过分子水平上的化学合成以及宏观复合材料的物理制备来实现高强度拉胀高分子材料的开发和应用[39~42]。可以相信这一天的到来为期不远了。

参 考 文 献

[01] Love A E H, *Mathematical Theory of Elasticity*, Dover, New York, 1944, 163; Voigt W, *Ann. Phys. Chem.* (Wiedemann), Vol. 31, 1887; Poisson S D, *Mem. Acad. Sci. Inst.* (France), 1829, 8:357

[02] Lakes R. *US Pat.* 4668557, 1987

[03] Lakes R. *Science*, 1987, 235:1038

......

[34] Evans K E. *Chem. Ind.*, 1990, 20:654

[35] Evans K E. *Endeavour*, 1991, 15:170

[36] Prall D, Lakes R S. *Int. J. Mech. Sci.*, 1997, 39:305

[37] Wei G Y, Edwards S F. *Physica*, 1998, A258:5; 1999, A264:388 & 404

[38] Wei G Y, Edwards S F. *Phys, Rev.* 1998, E58:6173

[39] Alderson A. *Chem. Ind.*, 1999, p.384

[40] 魏高原. 高分子通报,1995, p.234;当代化学前沿,1998, p.676;高分子辞典,1998, p.418

[41] 魏高原,吴红枚,曹维孝,冯新德. P & G 研究报告,2000 年 5 月

[42] Gaoyuan Wei, Hongmei Wu, Bingnan Jia. *International Symposium on Polymer Physics*, Huangshan, Sept. 13~17, 2000, Preprints, p.50

附：摘要及关键词的英译文

Auxetic Polymer Materials

Wei Gaoyuan

(*Department of Chemistry*, *Beijing University*, *Beijing* 100871)

Summary A few examples of specifically-fabricated auxetic polymers are briefly introduced, as are worldwide research activities on mechanisms that lead to auxetic behaviour of certain types of polymers, i. e., polymer auxeticity. Possible areas of application for auxetic polymers are pointed out and a perspective of their future is given.

Key words Auxetic polymer, Auxeticity, Auxetics

Translating (02) 化学反应的实时观察与 1999 年诺贝尔化学奖[1][2]
(Fast Probe of Chemical Reactions and 1999 Nobel Prize in Chemistry)

  1999 年度诺贝尔化学奖授予埃及出生并具有埃及和美国双重国籍的加州理工学院化学教授艾哈迈德·兹韦勒(Ahmed H. Zewail)，以表彰他在实现人类对化学反应的实时观察中所取得的先驱性贡献。瑞典皇家科学院的文告称，"兹韦勒教授所从事的研究使化学以及相关领域发生了一场革命——使人类能够理解并预料化学中的一些重要反应。"

  早在 19 世纪末，瑞典化学家也是现代物理化学的开创者之一的阿仑尼乌斯(Arrhenius)根据实验得出反应速度常数随温度变化的经验关系式，并首次提出"反应活化能"这一概念。在随后的 100 多年里，特别是随着量子力学的诞生，人类开始从分子层次上去认识化学反应，并于 1935 年提出了化学反应的过渡态理论。但由于过渡态的寿命很短，当时无法从实验上加以考察。随后，科学家应用闪光光解和弛豫方法研究了寿命为微秒(μs, 10^{-6} s)的反应中间体，并使用交叉分子束研究了高真空条件下分子反应的单次碰撞过程，使得人类对反应的认识更深入了一步。

  为了弄清反应的真实过程，人们必须了解分子内部的能量传递、反应物和生成物的能量状态，以及过渡态的真实状况。然而这些过程实际发生所花费的时间往往是在皮秒(ps, 10^{-12} s)和飞秒(fs, 10^{-15} s)的量级，因而，要实现对反应过程的检测就必须借助脉宽为飞秒的激光脉冲这一现代技术的结晶。在 1960 年第一台红宝石激光器出现之后的 20 年里，激光脉冲宽度从纳秒(ns, 10^{-9} s)逐渐缩短。1981 年出现了可产生 6 fs 超短激光脉冲的碰撞锁模染料激光器。20 世纪 90 年代以来，发展出更加稳定和易使用的全固体超快掺钛蓝宝石飞秒激光器，并逐渐取代了上述染料激光器。

  皮秒激光的出现使得对各种超快过程的研究成为可能。兹韦勒从 20 世纪 70 年代后期就开始利用超快激光研究化学反应，并自 80 年代起开始了一系列开创性工作，创立了飞秒化学(femtochemistry)这门化学物理学分支。在飞秒化学中，人们可以进一步了解发生在气相、液

---

  [1] 选自：蓝皮书《北京市科学与技术进步梗概》(2000 年，北京市科学技术委员会)第一部分的专题综述；作者：魏高原。

  [2] 本文引用了所列 4 篇参考文献的部分内容。

相、固相、团簇和界面中分子的动力学行为,可以帮助人们了解了发生在生物体系中的种种变化,同时,也为从量子态-态相互作用的层次上对化学反应过程实现控制提供了可能性。这无疑将对人类认识物质世界产生深远的影响。

那么,飞秒时间分辨是如何获得的呢? 显然,利用机械或电子学的方法是无法达到的。但是,由光的传播速度(3×10^8 m/s)可知,光在 1 fs 内只能走 0.3 μm。如果把飞秒激光分成两束,使之走过不同的距离,则两束光之间就会产生相对延迟。如果对光路进行非常精确的控制(精度达微米级),就可以产生可控制的飞秒量级的时间延迟。飞秒化学实验通常就采用两束飞秒脉冲激光。第一束激光用于启动化学反应,称做泵浦光;第二束光叫探测光。它经过不同延迟之后再作用于体系,相当于在反应启动后再在不同时刻给体系拍"快照"。从这些"快照"中可得到反应过程的演变信息。这种实验技术叫做泵浦-探测(pump-probe)技术。如果反应体系很复杂,可以应用同样的技术引入更多的飞秒激光脉冲,分别在不同时刻泵浦或探测。

人类首次直接观察一个化学反应过渡态变化过程是由兹韦勒于 1987 年实现的。在对氰化碘(ICN)进行的飞秒化学实验中,他第一次测得该光解反应(ICN→I+CN)的过渡态寿命约为 200 fs。在另一个重要实验中,兹韦勒研究了碘化钠(NaI)盐的光解反应(NaI→Na+I),第一次观察到了反应的过渡态在势能面上的振荡和解离的全过程。兹韦勒教授还研究了一系列从简单到复杂的化学和生物体系中各种类型的反应,包括单分子和双分子反应,其中有异构化、解离、电子转移、质子转移以及分子内部的弛豫过程,还有许多生物过程的反应。他在实验观察的基础上,也从理论上对这些过程进行了计算,并给予合理解释。他和同行在飞秒化学领域里的许多开创性研究成果大大推进了人类对化学反应微观过程在深度和广度上的认识和控制能力。

对化学反应进行实时观察的目的在于通过对化学反应规律的认识和掌握来实现控制化学反应—化学家长期以来梦寐以求的愿望。这种控制包括对反应产率和方向两方面的控制。前者如 NaI 的光解产率,可通过改变控制光与泵浦光之间的延迟来人为加以控制,后者则有利用超短激光脉冲诱导环酮的选键化学反应、过氧化氢(H_2O_2)和氘代水(HOD)的选键离解以及氢原子自氨(NH_3)至氘代氨(ND_3)的具有方向选择性的转移。

可以预料,人类对化学反应进行的实时观察最终将导致对化学反应特别是化学键的本质的认识,继而实现从宏观到微观对化学反应加以全面控制的梦想。而这一化学家的梦想的实现无疑又将通过在原子和分子水平上的各种"光加工"以制造食物和人类赖以生存的氧气等来造福人类,并最终使人类进行宇宙漫游成为可能。

参 考 文 献

[01] 孔繁敖,熊铁嘉,吴成印. 大学化学,2000,15(3): 5~8
[02] 兰峥岗,王鸿飞. 化学通报,2000,(1): 1~5
[03] 参考消息,1999 年 10 月 13 日,第 7 版;谢培,经济参考报,"开眼界"栏目,1999 年 11 月 5 日
[04] Service R F. *Science*,1999,286: 667~668

附录 E 科技会话常用课文与词汇
Appendix E Useful Texts and Vocabulary for Scientific Conversations

E.1 Selected Speaking Texts
Speaking Text (01) Plastics

The journalist, Giles Newton, and his wife, Susan, are at home after spending the day at a plastics exhibition.

SUSAN NEWTON I just can't help thinking of things made from plastics as imitations, as cheap substitutes.

GILES NEWTON If by "cheap" you mean less expensive, then you're quite right. For example, that new watering can we bought for the garden.

SUSAN Yes, it did cost less than a metal one.

GILES Do you remember why we bought it?

SUSAN I liked the nice bright yellow colour. But you can buy coloured metal ones, too.

GILES Ah! But with plastics, the colour goes all the way through, because the pigments are mixed in with the raw materials. They don't have to be painted like metal.

SUSAN What does it matter? The result's the same!

GILES No, it isn't. Take a watering can, or a child's toy, or even something you use in the kitchen, like your washing-up bowl. What happens when they're knocked against something hard?

SUSAN You mean if they're metal?

GILES Yes.

SUSAN I suppose, after a while the paint becomes chipped. All right, I see the point. With plastics the colours won't chip off.

GILES But do you remember another reason why we decided to buy a new watering can?

SUSAN Of course. The old one was so rusty. There were holes in the bottom...
    I see. Plastics don't rust like metal.

GILES Exactly. Are you beginning to feel more kindly towards plastics?

SUSAN I've nothing against them, Giles, but they are used instead of the original materials, so that makes them substitutes, doesn't it.

GILES Do you remember what Mr. Harvey said?

SUSAN Who?

GILES The plastics expert, you know, the chemist, in the recording I made at the exhibition.

SUSAN Oh yes, of course.

GILES And, incidentally, my tape recorder wouldn't be so small or so light if it weren't for the fact —

SUSAN I know, if it wasn't made of plastics.

GILES You're learning. I'll just run the tape back to the right place. I think this is where it is. Listen.

(*on tape recorder*)

GILES ... people who call them substitutes.

MR. HARVEY Oh, yes, some still do but they're quite wrong, Mr. Newton. Plastics are materials in their own right. Cheapness is not the only factor that makes them acceptable to industry. Before it can replace any other material—like wood, metal or a natural fabric—a plastics material must have a performance that is at least comparable to whatever was previously used.

SUSAN And I suppose sometimes they're even better.

MR. HARVEY Frequently, particularly when the properties of the material are adjusted, or even created, to suit the specific requirements of the end product.

GILES What sort of properties?

MR. HARVEY The degree of rigidity or flexibility, for example; resistance to acids, insulating qualities, ability to withstand sudden changes of temperature. Oh, the list is endless because the plastics industry is being asked continually to recommend or develop materials for such a wide variety of new uses.

GILES Do they succeed?

MR. HARVEY More often than not. In fact, there are so many types of plastics with so many unique properties, they frequently provide answers to unsolved engineering problems.

(*Giles Newton stops the tape recorder*)

GILES Well, Susan?

SUSAN He talks so easily about unsolved engineering problems. I'd be more impressed with an example—but a simple one, of course.

GILES As simple, perhaps, as your habit of leaving the refrigerator door partly open?

SUSAN Well, the catch is broken.

GILES Susan! It was repaired two months ago.

SUSAN Oh, all right! I sometimes give it a push with my elbow and it doesn't quite close. So?

GILES Well, somebody thought of making refrigerators without door-catches. Have you heard of polyvinylchloride—better known as PVC?

SUSAN Of course! The upholstery in the car, the kitchen floor tiles, the shower-curtains in our bathroom, they're all different types of PVC.

GILES Well, that's what was used to solve this particular engineering problem: PVC, with a magnetic filler.

SUSAN So, when the door is almost closed, magnetic attraction pulls it, keeps it tightly shut. That's very clever.

GILES And it's cheaper to make.

SUSAN And the refrigerator has a better door. Marvellous!

Speaking Text (02)　Oil

Giles Newton, and his wife, Susan, are visiting their friend, John Lane, an oil company executive. He's answering their questions about oil.

JOHN LANE　You see, Giles, I'm on the management side of the business.

GILES NEWTON　We only want a general background, John. And I know you've been to the oil wells and refineries.

JOHN　Just about all over the world.

SUSAN NEWTON　Then, please will you tell me the correct name for the stuff that actually comes out of the wells.

GILES　She means crude oil.

JOHN　That's petroleum.

SUSAN　Oh! Those are the sort of answers I always get! Which is it? Petroleum, or crude oil?

JOHN　Petroleum is crude oil, oil before it's refined. To be precise it's "crude petroleum oil", and we call it crude for short.

SUSAN　Oh! Oh, I see. This crude oil, this petroleum, what's it like?

JOHN　It varies from one oilfield to another.

GILES　Oh. Are there different kinds of petroleum?

JOHN　Well, they're all mixtures of hydrocarbons. But no two oilfields ever deliver crude of exactly the same composition.

SUSAN　John, I'll be quite happy with a general description: colour, for instance.

JOHN　The colour ranges from yellow through green to black. The lighter its colour, the easier it is to pour. The liquids vary in density from thin and treacly to thick and viscous. Sometimes it may be too thick to pour at all.

SUSAN　Oh, so it isn't always a liquid?

JOHN　Oh no, not always. It can be gas or vapour.

GILES　By the way, what exactly is bitumen?

JOHN　Well, the type of bitumen usually found on the earth's surface is a residue of petroleum from which all the more volatile elements have long since evaporated. It's black or brown, solid, perhaps, or a very viscous non-crystalline liquid.

GILES　The ancient Egyptians used bitumen, didn't they?

JOHN　Yes, five thousand years ago they used it as mortar, for water-proofing and even for road surfaces.

SUSAN　Bitumen must be one of the things you get when you refine crude oil.

JOHN　Yes. The first step in refining crude oil is fractionating by distillation. Look out of the window. Do you see those high columns?

GILES　Yes.

SUSAN　Oh, yes, those towers.

JOHN　They're distillation towers where we separate the oil into fractions. The crude oil is fed through a furnace into the bottom of the tower. We can't see inside the tower, so let me show you a diagram. Look, here.

SUSAN   Oh, I see.

JOHN   This fractionating results in six main raw materials.

GILES   I suppose these go through further refining and processing?

JOHN   Yes, and from this we derive hundreds of what you may call direct products.

GILES   And also basic ingredients for the development of thousands more?

JOHN   Yes. In the fractionating or distillation process, one of the six basic fractions is separated out as petrol, for use in cars, piston-engined aircraft and so on. This fraction is less than one-quarter of the crude oil that goes into the distillation tower.

GILES   What about all the rest of it, more than three-quarters?

JOHN   There's the gas that can be liquefied and bottled or piped into homes.

SUSAN   For cooking and heating?

JOHN   Right. There's the fraction called kerosine or paraffin.

GILES   For lamps and heaters?

JOHN   And for driving tractors.

SUSAN   Also jet aircraft.

JOHN   Yes, jet fuel is basically paraffin or kerosine, so is white spirit for paints —and also a great variety of insecticides. Do you want to hear more?

SUSAN   Yes, please! It's fascinating.

GILES   Doesn't another fraction of the crude oil produce diesel fuel for buses, trains and lorries?

JOHN   Yes. And another fraction is "de-waxed" to give us lubricating oils and greases for machinery of all kinds.

SUSAN   So they use everything. I'm sure they don't waste the wax.

JOHN   No, it has a lot of uses—for candles, waxed paper and many types of polish. Oh, Susan, this will interest you. Those same waxes are used to make cosmetics.

SUSAN   You mean the things I use for make-up? And the creams I rub into my skin?

JOHN   Yes, even your nail varnish is based on a by-product of crude oil. And some of the fractions of crude go through a process called cracking and the results provide a starting point for the synthesis of organic chemicals that nature is unable to supply in sufficient quantities.

GILES   I've read somewhere that oil-based chemicals are vital to the development of many plastics.

JOHN   Yes, including Susan's nylon stockings, also the soapless detergents she uses, shampoos, your car tyres, photographic film...

SUSAN   Please, John! Stop. My head's going round in circles.

JOHN   All right, but I warned you it could take three weeks to tell you the whole story!

Speaking Text (03) Computers

Giles Newton and his wife, Susan, are visiting a large computer installation in the head office of a group of warehouses dealing with engineering tools. They are talking to Henry Mitchell, an expert on computers.

MR. MITCHELL I'm afraid, Mrs. Newton, you're not quite accurate when you call a computer an "electronic brain".

SUSAN NEWTON Oh? But most people do.

MR. MITCHELL Yes, it was given that title by some misguided journalist not long after the first modern electronic computer was built in 1946.

GILES NEWTON It is electronic, Mr. Mitchell, so therefore you must be objecting to the word brain.

MR. MITCHELL Exactly, Mr. Newton. There are similarities with the human brain, but there is one very important difference.

SUSAN The computer is better?

MR. MITCHELL In some ways, yes; but not in something which is quite fundamental.

GILES Mr. Mitchell, do you mean the fact that the machine is controlled by man?

MR. MITCHELL I do. You see, despite all its accomplishments, the so-called electronic brain must be programmed by a human brain.

SUSAN Programmed?

GILES Yes, Susan, a program is a sequence of instructions prepared for the computer for a specific calculation, or series of calculations, enabling the computer to solve a given problem.

MR. MITCHELL The point is, that a human tells the machine what to do, when to do it and how it is done.

SUSAN I see! Do you also program information for the computer's memory?

MR. MITCHELL Correct, except the word memory has rather gone out of fashion. The computer doesn't really remember. Information is stored in it. We refer to core storage, or the store of data.

(*The door opens.*)

MR. BRINTON Mr. Mitchell, the machine's free now for a real time demonstration.

MR. MITCHELL Thank you, Jack. If you'd like to follow me... These are called interrogating typewriters.

GILES And that's the input-output system?

MR. BRINTON That's right. It's the method best suited to the particular needs of this organization.

MR. MITCHELL The operator—this young lady—types the information or questions, or both.

OPERATOR Yes. This produces two very different copies.

GILES One's an ordinary typewritten copy.

MR. MITCHELL   That's right. Then her typing also sends pulses down a line, an ordinary telephone line. And this simultaneously cuts the information or question on to a paper tape which is fed into the computer. I think Jack Brinton here can tell you the rest.

MR. BRINTON   Well, this is head office. We also have eight large warehouses scattered throughout the British Isles, each with a duplicate of what you see here.

SUSAN   Each with a computer?

MR. BRINTON   Oh, no, just one computer, and it's in this building.

SUSAN   And the eight warehouses are all connected to the same computer?

MR. BRINTON   Yes. Whenever any stock is received by or despatched from one of those warehouses, an operator sends the information online to the central processor. That gives us a complete running inventory. By the way, we supply all types of tools to the retail trade.

MR. MITCHELL   Mr. Newton, I thought it would make a good demonstration if you were to put any sort of question you like about the inventory to the computer.
Here it is.

GILES   Oh, well... How many six-inch screwdrivers are there in stock? Just the total for the eight warehouses.

MR. BRINTON   All right. Will you process that, please, Margaret?

OPERATOR   Yes, Mr. Brinton.

(*The operator types the question, and the reply is received immediately from the computer.*)

SUSAN   How soon will we get the answer?

MR. MITCHELL   It's there already, on the other machine.

SUSAN   It couldn't... It is! It says: "6" screwdrivers 14 gross."

MR. MITCHELL   That's known as real time processing.

GILES   So real time is a method of processing data so fast there's virtually no time-lag between enquiry and result.

MR. MITCHELL   Correct. For example, in the case of a man driving a car at, say, thirty miles per hour he can normally solve all his driving problems in real time; whereas at a hundred miles per hour he may not be able to.

SUSAN   I suppose, Mr. Mitchell, there must be many scientific uses for computers?

GILES   Weather forecasting, for example?

MR. MITCHELL   About a hundred years ago somebody had a theory for forecasting weather based on reports from all over the world. Unfortunately, it would have required sixty-four thousand skilled clerks working full-time.

GILES   I know about that. They use a similar system today, and one relatively small computer does the same work in only half-an-hour.

SUSAN   Is speed the only advantage?

MR. MITCHELL   It's one of them. Accuracy is another—provided the computer is given accurate information to start with.

SUSAN Can you give us another example?

MR. MITCHELL Well, in medicine computers help researchers to test drug by projecting information gained in limited trials. As a result, large-scale tests are not only becoming safer but also yielding far more useful information.

GILES Computers can also make mistakes, of course.

MR. MITCHELL Provided there is no hardware fault, the computer does not make a mistake. But its results are only as good as the information with which it is programmed. Let's go back to the office and we can talk about other scientific uses of computers. Come this way.

Speaking Text (04)　Lasers

Susan Newton turns off the television set.

GILES NEWTON Susan, I was enjoying that programme!

SUSAN NEWTON I want to know something, Giles, about laser beams. You've read the subject up, but I know nothing about it.

GILES But we're going to see a demonstration tomorrow.

SUSAN Yes, I know, but if I'm going along as your wife or your secretary—or both I'll be more of a credit to you if I know a little bit about it in advance.

GILES Okay! What do you want to know?

SUSAN Well, what is it? What does the word "laser" mean?

GILES Light Amplification by Stimulated Emission of Radiation.

SUSAN Oh yes, I see. L. A. S. E. R. laser. Quite simple, really!

GILES Once it's been explained to you, of course. Do you know anything at all about laser beams?

SUSAN Well, it was in that James Bond film "Goldfinger".

GILES Yes—I remember the scene. James Bond was lying on a gold table, and the table was being cut down the middle by a laser beam.

SUSAN The idea was that it would eventually cut James Bond in half as well. Could it really do that?

GILES Oh yes! At least, a laser could cut the table.

SUSAN But not James Bond?

GILES Don't be so disappointed. So far, if my facts are up-to-date, the laser won't penetrate the body.

SUSAN But I thought they were using it to cure cancer?

GILES Experimentally. And, even so, it can only be used on skin cancer.

SUSAN But what is a laser beam? You say it's a beam of light. But there are beams of light from car head-lights, from pocket torches, from searchlights and so on. They don't do any of the things that a laser beam does. So what is special about a laser beam? What makes it different from the other beams of light?

GILES Well, light consists of waves. These waves are very short—much too short to see directly. And ordinary light consists of waves all out of phase, out of step, with each other. White light or sunlight is also a mixture of every possible wavelength. Waves of red light are about twice as long as waves of blue light. So white light is a mixture of all possible wavelengths, all out of step with each other.

SUSAN   And what about laser beams?

GILES   Well, first all the waves in a laser beam have the same wavelength. A laser beam has a very definite colour. The red colour of the ruby is one of the most widely seen colours in them. But the difference between an ordinary beam of ruby red light and a laser beam of ruby red light is that in the laser beam the waves are also all in step with each other. So compared with any ordinary beam of light, the laser beam is a very orderly affair indeed. It's like a military march—everyone in step. In an ordinary beam, the waves are like the people in a crowd going to a football match, jostling and bumping into one another.

SUSAN   Does this orderly behaviour of the laser beam make a big difference?

GILES   Oh yes, and there's one more difference. Most beams of light, like the car headlamp, for example, are continuous. They shine all the time. But the laser beam is intermittent, and it's off much longer than it's on. Because these switches on and off are very fast, the eye doesn't see them. While the laser beam is off the energy for the next flash is building up, and when it comes, it's a very intense flash indeed. So lots of power can be packed into a laser beam.

SUSAN   Are there any other differences?

GILES   Yes. An ordinary beam of light diverges. It gets wider and wider, and therefore dimmer and weaker as it goes on. But a laser beam doesn't diverge in this way. So it carries its energy in a compact form, until it's absorbed when it strikes something opaque.

SUSAN   I see.

GILES   So that's why a laser can drill holes in hard jewels, for example, industrial diamonds, and others used in machine bearings.

SUSAN   Well, go on.

GILES   A laser can punch holes in steel, or in the tough ceramics used in the interiors of jet and rocket engines. It can be used like radar to track satellites—and to map cloud structures or...

SUSAN   That sounds marvellous. Anything else?

GILES   Well, there is one interesting point. In spite of everything it can do, the pulse from the largest laser ever made doesn't contain enough energy to boil a kettle for a couple of tea.

SUSAN   Is that some sort of a hint?

GILES   It did occur to me that, among your other talents, you can boil a kettle and make a cup of tea.

SUSAN   Well, since you've been so patient with me, Giles, I'll make you some. In fact, I'll even turn on the television for you again.

GILES   Thanks!

E.2 Useful Words for Scientifically Speaking
Word Study (01)

USEFUL WORDS

(1) *plastic*—permanently deformable by stress

　elastic—temporarily deformable by stress

All bodies are strained by stress: if the strain disappears after the stress is removed, the body is *elastic*; if the strain remains after the stress is removed, the body is *plastic*.

For an *elastic* material, Hooke's Law states that the ratio of strain to stress is constant. The constant is the coefficient of *elasticity* of the material for the particular type of stress.

Some materials, such as rubber, which are not normally *plastic*, can be *plasticized* by mixing them with substances called *plasticizers*.

Thermoplastic materials do not lose their *plasticity*. They can be heated and shaped as many times as required.

A modern synthetic polymer is distinguished from other plastic materials by being called a *plastics* material, not a *plastic* material.

Synthetic polymers such as PVC and polyethylene are called *plastics* materials; they are *plastic* only when heated.

Polyethylene is a *plastics* which is produced by polymerizing ethylene gas.

(2) *to synthesize*—to build from components

　to analyse —to identify or measure components

In the presence of sunlight, plants *synthesize* carbohydrates from carbon dioxide and water in a process known as *photosynthesis*.

Rubber was known only as a natural product until its *synthesis* was achieved in 1916. *Synthetic* rubber is now manufactured for a number of applications.

When the substance was *analysed*, it was found to be composed wholly of carbon, hydrogen and chlorine.

Analysis of the material by analytical chemists showed that it contained magnesium oxide.

(3) *to apply*—to put into position, to put into use

This adhesive must be *applied* to both surface which are to be bonded together.

The adhesive must not be allowed to dry after *application*.

Applying the law that action and reaction are equal and opposite, we see that the forces acting on a loaded beam are similar to the forces *applied* in breaking a stick.

Newton's third law of motion is *applicable* to this example.

For cheap, flexible containers, polyethylene is generally suitable but for some *applications* polypropylene has advantages.

　catalyst, to corrode, to design, to dissolve, to extrude, to form, granule, to impregnate, to insulate, to laminate, to package, to react, to resist, specific, to stabilize

WHAT THINGS ARE MADE OF, HOW THINGS PERFORM OR BEHAVE

(1) *element, compound, mixture, substance, material, structure, component*

Air is a *mixture* of the gaseous elements nitrogen (N) and oxygen (O), of gaseous *compounds such as carbon dioxide* (CO_2), and of water vapour.

Non-gaseous *substances* which occur naturally as pure elements, such as gold (Au), are rare and are often highly valued.

Viruses have a crystalline rather than a cellular *structure*.

The engine has more than 300 *components*, made of a number of different materials.

(2) *property, characteristic, feature, tendency, performance*

The physical *properties* of polyethylene include low specific gravity (S. G. 0. 92 to 0.96) and low tensile strength ($120 \sim 250$ kg/cm^2 at 23°C).

One of the *characteristics* of this plastics material is its *tendency* to become discoloured with age.

This aircraft has several interesting *features*.

Not all details of this aircraft's *performance* are available. But it is known to have a long range even at very high speeds.

(3) *rigidity, flexibility, brittleness*

Polystyrene is a rigid material. Its *rigidity* can be a disadvantage since it is also *brittle*. The brittleness of polystyrene can be decreased by the addition of rubber, to make it *flexible*. But the natural *flexibility* of many other plastics makes them preferable for certain uses, particularly for articles which must withstand frequent *flexing*. For example, the covering for the wire *flexes* of electrical goods is generally made of polyethylene or PVC.

Notice that a flexible cord is called a *flex*.

PREFIXES

Prefixes are most often added to words or roots of Greek or Latin origin so that most prefixes are themselves Greek or Latin. Such prefixes may also be put in front of English words. Occasionally a Greek prefix is found in front of a Latin word or vice versa (e. g., *television* formed from *tele*—at a distance, from Greek, and *vision*—vision, from Latin).

(1) *mono*—single

monochrome, monolayer, monomer, monomorphic, monoplane, monorail, monoxide

(2) *poly*—many

polycentric, polycyclic, polygon, polyhedron, polymer, polypeptide, polyvalent, polyamide, polybutadiene, polyester, polyethylene, polypropylene, polystyrene, polytetrafluoroethylene, polyurethane, polyvinylchloride

(3) *thermo-/thermo*—concerning heat

thermionic, thermocouple, thermodynamic, thermometer, thermoplastic, thermoscopic, thermosetting, thermostat. (N. B. the adjective *thermal vs.* the technical term *therm*, the latter is a unit of heat measurement.)

Word Study (02)

USEFUL WORDS

(1) *to extract*—to take out, to draw out

Waste gases can be *extracted* from an industrial process by mechanical *extractors*. The *extraction* is often done by means of fans.

Notice that in metallurgy to *extract* means to take the pure metal out of its ores.

Iron is *extracted* from iron ore by heating the ore with coke and limestone.

The *extraction* of aluminium is more complicated and is often performed electrically.

(2) *fluid*—flowing

A *fluid* substance is any substance which flows. Thus water and steam are *fluids*. Both have *fluidity*.

(3) *liquid*—neither gaseous nor solid

A substance is in its *liquid* state if it is neither gaseous nor solid. A gas can be *liquefied* if it is cooled sufficiently. For example, the *liquefaction* of the gas methane occurs at minus 160°C.

(4) *to melt* — to liquefy by heat

  to smelt—to extract iron from iron ore by melting the ore

At a certain temperature metals *melt*. They become *molten*. *Smelting* is the extraction of iron from iron ore by *melting* the iron ore. The ore is *smelted* in a blast furnace.

(5) *an oxide*—a compound of oxygen and another element

A compound of oxygen and another element is called *an oxide*, for example carbon *monoxide* (CO). When an element is caused to combine with oxygen it is *oxidized*.

Oxidation is extremely important in iron and steel making.

(6) *pure*—unmixed with other substances

A substance is *pure* if it is unmixed with other substances. *Impure* substances can be *purified*, and *purification* is regarded as complete when no trace of any *impurities* can be detected. In practice, it is difficult to *purify* any substance to 100% *purity*.

SOME WORDS TO DESCRIBE CONTENT AND COMPOSITION

(1) *to contain*

Pig iron may *contain* 4% of carbon. Its carbon *content* is 4%.

When all the *contents* are removed from a *container* (for example, the liquids from a vessel), the *container* is empty.

(2) *to consist of*

Carbon dioxide (CO_2) *consists of* carbon and oxygen.

(3) *to be composed of*

Steel is *composed of* iron and a number of other elements. Its *composition* can be varied to suit different uses.

A number of signals can be joined into a *composite* signal which can be transmitted and then *decomposed* into its several *components*.

COST

(1) *capital cost, initial cost, running cost, overhead costs, unit cost*

The *capital cost* of this steel-rolling plant was £25 000 000.

Though the *running costs* of this process are £1 000 per hour we must not forget also to add to the *total costs* the *initial cost* of heating the steel which is then rolled.

Because of the *overhead costs* we must run the plant continuously as far as possible in order to maximize output and so reduce *unit costs*.

PREFIXES

(1) *e-/ex*—out of

to eject, to emanate, to evaporate, to evolve, to exclude, to exploit, to extend, to extract, to extrude

(2) *im /in* into

to immerse, impact, to impel, to impinge, to impose, to impregnate, to induce, to inject, inlet, to install, to insert, to involve

re — again

to re-align, to re-connect, to re-enter, to re-heat, to re-process, to retract

N. B. the difference between the following pairs of words:

to re-count— to recount   to re-cover— to recover   to re-form—to reform

SUFFIXES

A number of technical terms are formed by adding a *suffix* to an existing word or word root. Her are some examples:

-ic	metal —metallic	atom —atomic	
	magnet—magnetic	electron—electronic	
-ar	nucleus—nuclear	line —linear	
-al	therm —thermal	electric—electrical	
	construction—constructional		
-ity	saline—salinity	active—activity	stable—stability
-ivity	conduct—conductivity		
-ation	saturate—saturation	rotate—rotation	

Word Study（03）

USEFUL WORDS

(1) *operate, operation, in operation, operational, operative*

It can *operate* only between airports with long runways.

(2) *increase* (noun), *increase* (verb), *decrease*

The air pressure is *increased* or *decreased* when the aircraft moves.

(3) *reduce, reduction*

Pressure above the wing is *reduced*.

(4) *act, act as, act on, active, activity, activate*

There are four forces which act on an aircraft as it files through the air.

(5) *design*—work out in detail

We have to *design* a short-haul aircraft for carrying freight.

The *design* of the new aircraft has just been started.

The *design* data will depend on tests of a new light alloy.

The *designers* have been asked to make several modifications to their original *design*.

(6) *designate*—name

The new aircraft type is *designated* "SH 603", but its commercial name will be "Pegasus".

The *designation* of the new aircraft type will be "SH 603".

(7) *modify*—change slightly, *specify*—state item by item

It is not too late to *modify* the shape of the tail fin to reduce drag.

Two *modifications* to the undercarriage have now been specified.

The complete *specification* must be drawn up by the end of this month.

The presence of an invading virus generates *specific* antibodies.

Every component of an aircraft has a *specifiable* function to perform.

(8) *behave*—act in a natural way, *perform*—act in a controlled way

The aircraft *behaved* well on its first test flight. It was immediately responsive to all the controls, all the specified tests were *performed* and repeated without trouble, and no unexpected faults were observed.

The learning *behaviour* of the white mice was studied by placing them in a complex maze and observing the number of trails they required to find the exit and the food without making any mistake. Once they had succeeded, subsequent trials tended to become confident *performances* of a routine carried out at high speed.

The department of *behavioural* sciences is making a special study of aggressive *behaviour* in animals and in man.

The new engines have now *performed* reliably for 2 000 h. Aircraft A has a better *performance* than aircraft B: its cruising speed is 50 km/h higher and its range is 120 km longer for the full load of fuel.

(9) *compare*—consider similarities and differences, *contrast*—consider differences

Compared with propeller-driven aircraft, jet aircraft have higher landing and take-off speeds and therefore need longer runways.

A *comparison* of Type A with Type B shows that Type A has more advantages for our purpose.

By comparison, Type A has the advantage of higher cruising speed and Type B the advantage of greater comfort.

In comparison with designers of propeller-driven aircraft, designers of jet aircraft have greater freedom in deciding where to place their engines.

The fuel consumption of a light petrol-engined two-seater aircraft is *comparable with* that of a sports car.

The huge bulk of the freight carrier *contrasts with* the slim streamlined form of the fast jet airliner.

The orange-yellow colour in which this aircraft is painted is intended to make as great a *contrast* as possible with the blue-grey sky background against which it operates so that it is clearly visible to other aircraft in the vicinity.

The model of the new supersonic aircraft shows advanced aerodynamic design; *by contrast*, even the most modern aircraft look clumsy and slow.

(10) *horizontal*—parallel to the horizon, level
   vertical—perpendicular to the horizon, upright

An ideal projectile is a particle which moves in a vacuum with constant *horizontal* velocity under the action of a constant *vertical* force.

A spirit-level shows the *horizontal*

A steady plumb-line shows the *vertical*

Aircrafts take off *horizontally*; helicopters take off *vertically*.

To an observer at elevation h above the surface of the Earth, the distance of the horizon is approximately $(2hR)^{1/2}$, where R is the radius of the Earth.

The area of a triangle is $bh/2$, where b is the length of the base and h is the height of the vertex above the base.

(11) *obsolete*—out of date

Supersonic aircraft will not make all subsonic aircraft *obsolete*; many jobs remain that only subsonic aircraft can do.

The development of aircraft design is now so rapid that by the time a new aircraft has been designed, tested and certified as airworthy it is already *obsolescent*.

The period of *obsolescence* varies from one aircraft type to another; in the case of a modern airliner, it is assumed that the aircraft will have become *obsolete* after eight years.

(12) *nucleus, nuclear, nuclear fission, fissile*

Nuclear fission takes place when a free neutron strikes the *nucleus* of *fissile* element such as uranium.

(13) *react, reactor, reaction*

The heavy fractions are vaporized and enter a *reactor* where...

Platinum is used as the catalyst in a reaction which takes pace in a platformer.

A virus breaks down the carbohydrate constituents of the cell wall by enzymic *reaction*.

Ammonia *reacts* very readily with sulphuric acid to form ammonium sulphate.

As with the two-*reactor* station, all plant in the single *reactor* station is housed in one composite building.

A chain *reaction* is started. To slow down the nuclear fission, boron steel rods are dropped into the *reactor*.

Action and *reaction* are equal and opposite.

Each monomer molecule can be made to *react* with similar monomer molecules to form a chain. This *reaction* is known as polymerization.

The energy released by the *nuclear reaction* heats a stream of gas.

Urea is made by combining ammonia with carbon dioxide but this *reaction* occurs only...

(14) *generate, generator*

There is one turbo *generator* for each reactor.

The hot gas is blown over thousands of tubes of water, thus *generating* steam which drives ordinary turbine *generators*.

A nuclear reactor *generates* neutrons which react with the fissile materials of the core.

(15) *condense, condenser, condensate*

When steam impinges on a cold surface it *condenses* to form water.

A *condenser* is a device which exposes a large area of cold surface to any steam which enters it.

Distilled water is the *condensate* from the steam.

(16) *control(verb), control(noun used as adjective), fuel, refuel*

The fission process is *controlled* at the desired level of activity by moving *control* rods.

The boilers of conventional power stations have to be charged with oil or coal continuously, but nuclear power stations are *refuelled* only at long intervals.

(17) *vapour*—the gaseous state below the critical temperature

　　　evaporate—convert from liquid to vapour

The amount of water *vapour* in the atmosphere is usually measured by a hygrometer.

At its boiling point a liquid has a *vapour* pressure equal to the external pressure.

Petrol *vaporizes* readily at normal temperatures.

The latent heat of *vaporization* of water is 537 calories per gram at 100°C and 76 cm Hg pressure.

If pure water is *evaporated* to dryness, no residue remains.

The *evaporation* of water from reservoirs is reduced by allowing a thin film of oil to spread over the surface.

(18) *condense*—physical process—convert vapour to liquid or solid by cooling

At normal pressures steam *condenses* at 100°C.

If air is cooled, the temperature at which *condensation* of water vapour first occurs is called the dew point.

The *condensate* obtained from a boiling mixture of water and alcohol also contains both water and alcohol.

A well-designed *condenser* exposes within a small volume a large area of cooled surface to the incoming vapour.

(19) *distil*—vaporize in one vessel and collect the condensate in a second vessel

Heavy oil fractions have to be *distilled* under reduced pressure.

Distilled water is tasteless and odourless.

Water and alcohol cannot be separated by *distillation*.

The process of *distillation* separates a liquid into a *distillate* and a residue.

(20) *lubricate*—make run smoothly

Metal-to-metal surfaces are *lubricated* to reduce friction.

Lubricating oil reduces wear between metal surfaces by separating the surfaces.

Lubrication reduces noise and wear.

Oil is not the only common *lubricant*; graphite is also commonly used.

(21) *proportion*—numerical relation of a part to the whole

　　　ratio—numerical relation of one part to another part

Hydrogen and oxygen are present in water in the *proportions* of 1 to 8 by weight.

The areas of triangles standing on the same base are *proportional* to their heights.

The profits of the enterprise will be divided *proportionally* among the participants on the basis of the work each one does.

One gram of A reacts with three grams of B, and so on *proportionately*. For other quantities of A proportionate amounts of B are required.

Compared with the human adult, the new-born human baby has a *disproportionately* large head.

If A∶B=C∶D, the *ratio* of A to B equals the *ratio* of C to D.

If two maps of the same area have different scales in which distances are in the *ratio* of 1∶5, then comparable areas are in the *ratio* of 1∶25.

(22) *catalyst*—a substance which increases the rate of a chemical reaction but which itself remains unchanged

In the presence of mercuric sulphate as *catalyst*, acetylene combines with water to form acetaldehyde:

$$C_2H_2 + H_2O \longrightarrow CH_3CHO$$

The reaction in which ammonia mixed with oxygen is oxidized to nitric acid is *catalysed* by platinum wire grids.

Water has *catalytic* properties difficult to demonstrate because, in any chemical system, the last traces of water can be eliminated only after prolonged treatment.

Catalysis of a second degree, in which the activity of a catalyst is increased by a second substance, is also known.

Word Study (04)

USEFUL WORDS

(1) *to construct*—to put together, to build

It takes about two years to *construct* a large bridge.

The *construction* of a large bridge takes about two years.

The bridge would be *under construction* for two years.

Suspension bridges are usually *constructed* over water.

Brunel was a famous *constructor* of bridges and tunnels.

If a bridge collapses, it is important to find out whether the cause was a constructional fault (fault in construction) or an error in design.

(2) *to compress*—to press together

The forces in the upper part of a loaded beam are *compressive*.

The upper part of a loaded beam is in *compression*.

All permanent gases are equally *compressible*.

Solids are relatively *incompressible*.

Compressed air is supplied by a *compressor* (a mechanical pump).

(3) *to tension*—to tighten (a string, cable or tie-rod)

In casting pre-stressed pre-tensioned concrete, it is necessary *to tension* the reinforcing wires before the concrete is cast.

The wires must be *put in tension* before the concrete has set.

The lower part of a loaded beam is *in tension*.

The forces in a cable can only be *tensile*; they cannot be compressive.

(4) *to suspend*—to hang, to stop temporarily (suspension—a fluid mixture of liquid and solid particles)

All the cables and *suspenders* being in position, the next task is to *suspend* the deck section by section.

A *suspension* bridge is used for very long spans.

When ice forms on the bridge, construction work has to be *suspended*.

At what wind-speed has it been decided to *suspend* operations?

When carbon dioxide is bubbled through clear limewater, a milky white *suspension* of calcium carbonate is produced.

(5) *to maximize*—to make as large as possible (maximum, maxima)

   to minimize—to make as small as possible (minimum, minima)

The streamlined cross-section *minimizes* wind effects.

What is the *maximum* height of the deck above water-level?

(Notice that in science words of Latin origin are preferred to words of Anglo-Saxon origin, e. g. : the scientist prefers *suspend* to *hang*, *construct* to *put together*, *compress* to *press together*.)

(6) *horizon*, *horizontal*

The distance of the *horizon* at sea is proportional to $h^{1/2}$ where h is the height of the observer above seal-level.

To test whether a surface is *horizontal* it is necessary to apply a spirit-level in two perpendicular directions.

(7) *vertex*, *vertical*

The *vertex* of a triangle is the corner opposite the line that is selected as the base line.

To test whether a surface is *vertical* it is usual to apply a plumb-line.

(8) *distinguish*, *distinction*, *distinct*

How is civil engineering *distinguished* from other branches of engineering?

The engineering profession now has many *distinct* branches.

(9) *reduce*, *reduction*

The average lengths of the current-paths will be *reduced* to one tenth.

The main effect of the atmosphere is the *reduciton* of the violent changes of temperature.

(10) *develop*, *development*

Sometimes a computer takes so much power that cooling systems which require still more power have to be installed to keep it from getting too hot, which would increase the risk of faults *developing*.

The *development* of miniaturized computers has been important in the aircraft industry.

(11) *electron, electronic, electronics, electric, electricity, electrical*
The *electron* is a sub-atomic particle with a negative charge.
At this university it is possible to study *electrical* engineering and *electronics*.
The television receiver is an *electronic* device.
He is an *electrical* engineer who has specialized in electric traction.
(12) *automate, automatic, automated, automation*
Welding can be *automated* in a variety of ways.
There are ships at sea with *automatic* helmsmen.
The gyro compass *automatically* keeps the ship on a predetermined course.
Automation at sea requires controls that operate complex interacting systems.
The shipboard computer would *automatically* alter course.
(13) *propel, propeller, propulsion, propulsive*
Speeds of rotation of different shafts, including the *propeller* shaft, are measured.
These factors must be adjusted rapidly to ensure maximum *propulsive* efficiency.
There are various possible alternatives for ship *propulsion* machinery.
The horse-power needed to *propel* the ship must be calculated.
(14) *power* (noun), *power* (verb)
The *power* needs of the ship must be ascertained.
In the *power* range 3 000 to 10 000 *horse-power* diesel has advantages.
The ship can be *powered* by any well-tried engine.
The ship will be *powered* by steam turbines.
(15) *control* (noun), *control* (verb)
Real automation at sea requires *controls* that operate complex interacting systems.
The movement of the cutting head is *controlled* either numerically by computer-tape or optically.
The *control* head follows a specially-prepared scale drawing.
(16) *stable, stability, stabilize*
The decision in regard to dimensions will depend on the *stability* of the hull.
The ship's *stabilizers* are driven by electric motors.
(17) *to implement, implementation*
We cannot *implement* our computer program until the tests on the computer peripherals have been completed.
Now that the plans for installing the computer have been approved, we start at once on their *implementation*.
(18) *to compute, to computerize*
Here is a program for *computing* the value of π to as many decimal places as you need.
We are planning to *computerize* all the clerical pay-roll routines.
(19) *to affect, to effect, effect, effective*
The *effect* of the modification was negligible.
The modification has not *affected* the result.

Though the modification has been *effected* it has been *ineffective*.

PREFIXES

Notice these prefixes:

$$super- + \text{structure} \longrightarrow \text{superstructure}$$
$$pre- + \text{fabric} \longrightarrow \text{prefabricate or prefabrication}$$
$$pre- + \text{determine(d)} \longrightarrow \text{predetermine(d)}$$

Word Study (05)

USEFUL WORDS

(1) *to vary, varied, variations, variety, various, variable*

Variations in the temperature, pressure and humidity of the air through which the laser beam passes cause *variations* in the wavelength.

Daily and seasonal *variations* in the solar radiations lead to *variations* in atmospheric temperature and pressure.

The pattern of winds is therefore continuously *variable*.

Plastics are manufactured in a *variety* of forms.

The D. L. I. S. receives regular reports in *various* countries.

Engineers use *various* methods to reduce fluctuation in demand.

Most *varieties* of steel contain elements such as carbon, silicon and phosphorus.

(2) *to measure, measuring, measured in, measurement, measurable*

A simple piece of apparatus delivers a *measured* volume of insecticide on to the locust.

Lasers can be used as *measuring* instruments for making accurate measurements of very small objects or displacements.

The enormous distances of galactic space are usually *measured* in terms of the light-year which is the distance light travels in one year.

(3) *to impinge, impact*

According to the law of momentum, when two elastic particles *impinge* the total momentum before *impact* equals the total momentum after impact.

(4) *to emit—emission; to transmit—transmission; to absorb—absorption*

The light *emitted* from the glowing filament of an electric lamp has a wide range of wavelengths; it is polychromatic.

The *emission* of light from a laser occurs at one definite wavelength characteristic of the laser; it is monochromatic.

Light is *transmitted* by transparent materials such as glass but is either absorbed or reflected by opaque materials

(5) *to excite—excitation*

When one or more of its electrons jumps to a higher energy level, an atom is said to be *excited*.

The energy of *excitation* of an atom is released by the emission of light.

(6) *to phosphoresce—phosphorescent; to fluoresce—fluorescent*

Phosphorescent materials glow in the dark because chemical energy within them is being converted directly into light. Certain fishes and insects *phosphoresce* in this way.

In *fluorescent* materials energy previously absorbed is emitted again as light.

Word Study (06)

USEFUL WORDS

(1) *finite, infinite, definite, indefinite*

The binomial expansion of $(q+p)^n$ is *finite* when n is an integer; the number of terms in the expansion is $(n+1)$.

The numerical sequence 1,2,3,4... can be continued *indefinitely*, without limit; such a sequence is *infinite*.

As the value of x tends to c, the value of the expression $(x-c)/(x^2-c^2)$ tends to $0/0$, which is *indefinite*. But if the common factor $(x-c)$ is first cancelled, the fraction reduces to $1/(x+c)$ and its value tends to the *definite* value $1/(2c)$ as x tends to 1.

To a pure mathematician concerned always with exact *definite* quantities, the probabilistic estimates of the statistician always look very *indefinite*.

(2) *continuous, discrete*

Statistical variables may be either *continuous* or *discrete*; *continuous* variables, such as lengths, are usually measured; *discrete* variables, such as numbers of people, are usually counted.

(3) *frequent, frequency, frequently*

The *frequencies* of publication of scientific journals vary widely; some are published weekly, some monthly and some only four times in the year.

The *frequency* of the alternating current supplied to this radio set must be 50 cycles per second.

The set of statistical tables in my office is needed for *frequent* reference.

When I am tired I *frequently* make silly arithmetical mistakes.

(4) *probable, probability, possible, possibility*

The *probability* that a tossed coin shows "heads" in n successive throws is $(1/2)^n$.

It is *possible* that all the telephone subscribers on this exchange will try to telephone simultaneously but, fortunately, it is highly *improbable*.

It is incorrect to say that perpetual motion without absorption of energy from some source is *improbable*; it is *impossible*.

(5) *to distribute—distribution; to disperse—dispersion; to deviate—deviation*

In a lottery the probability of success should be *distributed* uniformly over all legitimate possibilities.

The heights of adult men are found to conform with the normal or gaussian *distribution*.

The standard *deviation* is a measure of the degree of *dispersion* of a variable about its mean value. The more widely the variable is *dispersed*, the greater is its standard *deviation*.

(6) *to amplify*—to make larger

This *amplifier* has a gain of about 10.

The *amplification* of the signal must be achieved without distortion.

(7) *to compose*—to put together, assemble, *composition*, *composite*

The component parts of the *composite* signal have to be separated again.

(8) *to extend*—to stretch out; *extensive*, *extensible*, (*extent*)

The aerial normally has a length of 1.5 metres but is *extensible* to 2.5 metres.

The range of the local relay station is being *extended* from 10 to 15 km.

The *extent* of the damage is being assessed.

(9) *to align*—to put in line

The horn aerials are *aligned* along the line of sight.

If they are in *alignment*, signals can be received; if they are out of *alignment* they cannot.

(10) *to multiply*—*multiplication*, *multiple*, *multiplicity*

The result of *multiplying* 3 by 7 is 21.

The *multiplication* of 3 by 7 yields 21.

The smallest *multiple* of 3 and 7 is 21.

The separate insulated wires are wound together to form a cable with a *multiple* core.

A single frequency band can accommodate a *multiplicity* of operating channels.

(11) *to deflect*—*deflection*; *to reflect*—*reflection*; *to refract*—*refraction*

The windscreen of a moving car *deflects* the air-flow over the driver's head.

When one looks at a *reflection* in a plane mirror, the image appears to be as far behind the mirror as the object is in front.

Light is *refracted* when it passes from one medium into another of different density. The angles of incidence and of *refraction* are related by Snell's law.

The gravitational field of the sun causes a *deflection* of any beam of light which passes near the sun.

附录 F 基本化学术语总汇及索引
Appendix F Baisc Chemical Terms and Index

absolute deviation 99
absolute entropy 95
absolute error 99
absolute temperature scale 27
absolute zero 27,95
absorption of radiation 102
absorption 83,98,102~104
acceptor atom 77~78
accuracy 10,99~100,157
acetylenes, alkynes 46~47
acid (water-ion) 82
acid anhydride 49~50
acid or base ionization constant 84
acid salt 83
acidic aqueous solution 82
acidic anhydride (oxide) 83
actinides 77
activation energy 93
actual mass of an atom 16
acyl group 50
acyl halide 50
addition reactions 86
adenosine diphosphate (ADP) 117
adenosine triphosphate (ATP) 117
adsorption 93~94
aerosol 93,131,175
alanine (Ala) 117
albumin 118
alcohol 25,27~29,48
aldehyde 48~49,52,87
aliphatic (hydrocarbons) 86
alkali metals 75,77
alkaline aqueous solution 82
alkaline earth metals 77
alkanes 46~47,86~88
alkenes (olefins) 46,86
alkyl groups 50~51,86

alkyl halide 48
alkynes (acetylenes) 46,86
allotropes 83
alloy 7,156,186
alternating copolymer 109
amide 50
amines 48~49,88
amino acid 51~52,56,58,117
amorphous solid 91
amphoteric oxide 83
anabolism 117,142
analytical chemistry 5,97~98
angular momentum 76
anions 20,24,78,82
anode 95~96,102
antineutrino 81
aqueous solution 17,22,33~35
arginine (Arg) 117
aromatic hydrocarbons 47,86
aryl halide 48
aspartic acid (Asp) 117
asymmetric atom 86
atactic polymer 109
atomic mass unit 16,22
atomic mass 16,18,22,25,44
atomic number 16,75,81
atomic orbital 76,79,161
atomic radii 78
atom 15~19
Avogadro's law 27~28
Avogadro's number 22,25
back-titration 100
barometer 103
base (water-ion) 82
basic anhydride (oxide) 83
bidentate 84
bimolecular reaction 93~94

binary compound   20,24
binding energy per nucleon   80
biological sciences   3,6,8,97,184
block copolymer   109,179
bombardment reactions   81
bond angle, bond axis   79
bond dissociation energy   78
bond energy   78,161
bond length   78
bond order   84
bonding or antibonding molecular orbital   84
Boyle's law   27~28,30
breeder reactor   81
Brønsted-Lowry acid   83
Brønsted-Lowry acid-base reaction   83
Brønsted-Lowry base   83
buffer solution   94~95
burets   101
calibration   99,101
calorimeter   43~44,103
carbohydrate   52,61,63~64,116
carbonate hardness, temporary hardness   82
carbonyl group   48
carboxylate ion   49
carboxylic acid   49~50,56,58,63
catabolism   116,137,142
catalyst   22,35,38,55,62
catenation   83
cathode rays   15~16
cathode   95~96
cations   20,24,78
cell potential, electromotive
　　force or emf   96
cell reaction   95
cellulose   52,54~55,63,106
chain reaction polymerization   106
chain reaction   81,106
changes of state   17,43
Charles' law   27~28
chelate ring   84

chelation   84
chemical analysis   97~98,105,120
chemical bond   20,40,77,94,105
chemical change   3~6,8~9,21
chemical compound   4,17,19~20,22~23
chemical equation   21~22,24,32,35
chemical equilibrium   82
chemical formula   19~20,68,71,88
chemical kinetics   93
chemical nomenclature   20
chemical properties   17,110,119
chemical reaction   17,21,24~25,27
chemical reactivity   77
chemical stability   77
chemistry   3,8
chirality   86
chromatography   104,142,162
chromatograph   162
collagen   117~118
colligative properties   92
colloid   93,111
colloidal dispersion   93
colorimetric analysis   98
combination reaction   33,38
combustion   6,37~38,41,44
common amino acids   117
common ion (effect)   94
complex ion   82,95
complexometric titration   101
complex   84~85
concentration   23,82~84,92
condensation, liquefaction   90
conductometric titration   102
configurational unit   109
configuration   76,79,116,119
conformation   119~121
conjugated double bonds   86
constant   26
constitutional repeating unit   109
constitutional unit   109

附录 F 基本化学术语总汇及索引(Baisc Chemical Terms and Index)

coordinate covalent bond 78
coordination compound 84
coordination number (complex) 84
coordination number (crystal) 91
copolymer 51,106,109,179
coulometric titration 102
couple action 96
covalent bonding 77
critical mass 81
critical point 90
critical pressure 90
critical temperature 90
crucible 101,157
crystal structure 91
crystalline soild 91
crystal 91,104,111,119
cycloalkanes 46~47,86
cysteine (Cys) 117
Dalton's law of partial pressures 29
decomposition reaction 33,38
degree of polymerization 109
deliquescent 82
delocalized electrons 80
denaturation 118
density 12~13,18,28,31
deoxyribonucleic acid (DNA) 116,136~137, 140,142
descriptive chemistry 5,8
desiccator 101
deuterium 83,104,159
diamagnetism 76
diatomic molecule 19
diffraction 75,104,111~112,121
diffusion 29,135
dipole moment 80
dipole 78,80
dipole-dipole interaction 80
displacement reaction 33,36,38
disproportionation reaction 83
dissociation constants (complex ions) 95

dissociation of an ionic comtound 22
distillation 87,99
donor atom 77~78,84
effective nuclear charge 78
efflorescence 82
effusion 29~30
elastomers 107,175
electroanalysis 98
electroanalytical chemistry 101
electrochemical cell 95,101
electrochemistry 95
electrode reaction 95
electrode 15~16,95~96,102
electrogravimetric method 101
electrolytic cell 95
electrolysis 95
electromagnetic radiation and spectra 103
electron affinity 79
electron capture 81
electron configuration 76,79
electron spectroscopy for chemical analysis (ESCA) 105
electronegative atom 79~80
electronegativity 79
electronic structure 75,161
electron 15~16,76~78
electrophile 87
electropositive atom 79
elementary reaction 93~94
element 16~17,19~20,22
electrolytes 82
emission of radiation 103
empirical formula 23~25,31
empirical relationship 44
emulsion 93,178
enantiomers, optical isomers 86
end point 95,100~102
endothermic 40~41
enthalpy 40~44
entropy 95,107,184

enzyme   55,63,116~117,123
equilibrium constant   94~95
equivalence point   100~101
equivalent mass of an acid or a base   83
ester   48~50,87~88
ether   48~49,87~88
evaporation   43,90,128,151
exact numbers   10
excited state   76,105
exothermic   40~41,45
faraday   96
ferromagnetic   186
fibrinogen   118
first law of thermodynamics   40
flux   129
foam   71,93,115,169
formula unit   20,25
fossil fuels   82,86,127,129~130
free energy change   95
free radical   93,111
frequency   12,62,75,103~104
fuel cells   96
functional group   48~49,53,56,87
fundamental particle   16
fusion   43,80,90
gas-discharge tube   15~16
Gay-Lussac's law of combining volumes   27,30
gel   93,118
giant molecule   109
glass transition temperature   106~107
glutamic acid (Glu)   117
glycine (Gly)   117
graft copolymer   109
Graham's law of diffusion   29
Graham's law of effusion   29
gravimetric analysis   98,101
gravimetric method   99,101
ground state   76,81,105
group, family (periodic table)   76

half-cell   95~96
half-life   81,85
half-reaction   95~96
halogen   77,86
hard water   82
heat capacity   43~44
heat of reaction   41~42,45
heat   6,9,13,34,40
heavy water   83
Heisenberg uncertainty principle   75
Henry's law   92
Hess's law   42
heterogeneous mixture   17
heteronuclear   84
halogenation   86
hexagonal or cubic closest packing   91
high (molecular weight) polymer   109
homogeneous mixture   17
homogeneous or heterogeneous reaction   94
homogeneous or heterogeneous catalyst   94
homologous series   47
homonuclear   84
homopolymer   106,109
Hund's principle   76
hybridization   79
hydrates   82
hydration   82
hydrogen bond   80,118
hydrogenation   62,111
hydrolysis of an ion   94
hydrolysis   82,94,108,117,122
hydrophilic   116
hydrophobic   116,140,144
hydroxyl group   106
hygroscopic   82
ideal gas law   28,30~31
ideal gas   18,26,28,30~31
ideal solution of a molecular or an ionic solute   92
immiscible   92

indicator 95,100,102
infinitely miscible 92
infrared (IR) spectroscopy 104
inhibitors 94
initial reaction rate 94
inorganic chemistry 4~5,8
instrumental analysis 98
intact DNA 118
interference 99
intermediate 37,81,93,137,141
intermolecular forces 80,92~93,106,109
internal energy change 40~41
internal energy 40~41
internal redox reaction 83
interpenetrating polymer network (IPN) 109
ion exchange 82
ion product constant for water 84
ion product 84,95
ionic bonding 77~78
ionic radii 78
ionization energy 79
ionization 22,79,84,105
ionomer 109
ions 16,18~20,22
ion-electron equation 95
irregular polymer 109
isoelectronic ions 78~79
isoleucine (Ile) 117
isomers 50,52,80,86,88
isomorphous 91
isotactic polymer 109
isotopes 16~18
Kelvin temperature scale 27
keratin 118
ketone 49,53,116,164
labile or inert complex 84
lanthanide contraction 78~79
lanthanides 77
large assemblies 118
laser spectroscopy 105

lattice energy 78
Le Chatelier's principle 83
leucine (Leu) 117
Lewis acid 84
Lewis acid-base reaction 84
Lewis base 84
Lewis structures 77~78,88
Lewis symbol 77
ligands 82,84,175
limiting reactant 36,42
lipid 116,140
London forces 80
lysine (Lys) 117
macromolecules (giant molecules) 109
magic numbers 81
manometer 103
mass defect 80
mass number 16,18,81,83
mass percent 92
mass spectrometer 103,105,162
mass 8,10,12~13,16
matter 3~5,8
mean 98
median 98
relative deviation 99
molecular spectroscopy 103
melting point 69,90,166
metabolism 117,123,135,139,142
metallic bonding 77
metals 5,33~34,39,77
methionine (Met) 117
mineral 23,97,101,126,174
miscibility 92
mixture 4,6,8~9,17
molality 92
molar heat capacity 43
molar mass 22~24,28,30
molarity 23,25,35,92,162
mole fraction 29,31,92
molecular formula 23~24,28,31,47

molecular geometry 79
molecular mass 22,30,35,48,87
molecular orbital (theory) 79,84
molecularity 93
molecule 18~19,21~23
mole 12,22~25
momentum 75~76
monomer 109
multiple (double, triple) covalent bond 77~78
multiple unit cell 91
multiple-stranded helices 120
natural rubber 107,176
net ionic equation 34~35,38
network covalent substance 78
neutralization titration 101
neutralization 82,101
neutrino 81
neutron number 16
neutron 15~16,18,80~81
noble gas configuration 79
noble gas 76~77,79
nonbonding electron pairs, lone pairs 77
non-carbonate hardness, permanent hardness 82
nonelectrolytes 82
nonmetals 77~79
nonpolar covalent bond 78
non-stoichiometric compound 91~92
normal boiling point 90
normal freezing point 90
normal hydrocarbons 46
normality 83~84
nuclear binding energy 80
nuclear chemistry 80
nuclear fission 80~81
nuclear force 80
nuclear fusion 80
nuclear magnetic resonance (NMR) spectroscopy 104,159

nuclear reactions 81
nuclear reactor 81
nucleic acid 116,118~121
nucleons 80
nucleophile 87
nucleotide 116,119~120,122,125
nucleus 15~16,80~81,83
nuclide 80~81
octet rule 77
olefins (alkenes) 46
oligomer (having 2~10 monomer units only, low molecular weight polymer) 109
optical isomer 86
ore 15,37
organic chemistry 4,8,52,88~89
osmosis 93
osmotic pressure 93,112
overall reaction order 94
overvoltage 96
oxidation number, oxidation state 79
oxidation-reduction titration 101
oxidation 49,79,82~83,95
oxidation-reduction reactions 83,95~96
oxidizing acids 83
oxidizing agent 82~83
oxidizing anion 83
oxo acid 83
paramagnetism 76
partial pressure 29~31,92,96
partner-exchange reaction 33,38
Pauli exclusion principle 76
peptide bond 58,63,65,67,116
percent yield 36~37,39
percentage composition 23,25,159
periodic table 75~77
period 76~77,79
petroleum alkylation 87
petroleum cracking 87
petroleum isomerization 87
petroleum reforming 87

pH and pOH 83~84
phase 40,43,90,92,94
phenol 48
phenylalanine (Phe) 117
photoelectric effect 75
photon 75,105
physical change 4,8,41,43,89
physical or chemical adsorption 94
physical properties 9,17,106,112
physical sciences 3,8
pipets 101
plane-polarized light 86
polar covalent bond 78
polarization of an ion 79
polyamide 67,108,112,117,120
polyatomic ions 20,24
polyatomic molecule 19
polymer 5,54,56,63,106,109
polymorphous 91
polynucleotide 123
polypeptide 58~59,64,67,116
polyprotic acids 82
polysaccharide 52,54~55,63~64
polystyrene 51,71,108,111,115
positron 81
potentiometric titration 102
precipitation method 99
precipitation titration 101
precipitate 33~34,36,82,99
precipitation 34,99,101~102,128
precision 10,99,101
pressure 12~13,16,22,26
primary standard 100
primary structure (sequential order of the residues) 119
primitive unit cell 91
principles of chemistry 5,8~9
products 21~22,27
promoters 94
protein 52,56,58~60

proton 15~16,50,80~81
pseudo-noble gas configuration 79
pure substance 17,19,82,95
qualitative analysis 5,8,98
quantitative analysis 5,8,98,104
quantized 75
quantum mechanics 75,183
quantum number 76
quantum theory 75
quantum 75~76,161,183
quaternary structure (the arrangement of subunits) 121
racemic mixture 86
radioactivity 81
radionuclides 81
radius ratio 91
random copolymer 109
Raoult's law 92
rare earth elements 77
rate constant 94,135
rate equation 94
rate-determining step 94
reactant reaction order 94
reactants 21~22,32~33,35
reaction mechanism 93~94,116
reaction quotient 94~95
reaction rate 94,138~139
redox couple 95
redox reactions 83
reducing agent 82~83
reduction 82~83,95
regular polymer 109
relative error 99
representative elements 77
resin 108,177
resolution 118,121,162
resonance 47,78,104,138,159
resonance hybrid 78
ribonucleic acid (RNA) 116,136,143
salts 20,24,49,131

saturated hydrocarbons　46,86
saturated solution　92
screening effect　78
secondary structure (helices of residues)　120
semiconducting elements　77
semipermeable membranes　93
serine (Ser)　117
significant figures　10～11,14,17
simplest formula　23,28
single covalent bond　77～78,86
solubility product　95
solubility　92
solute　17,23,84,92,104
solution　9,17,22～23,25
solvation　82,112
solvent　17,23,48,82,92
sol　93
space lattice　91
specific heat　43～44
spectator ions　34
spectrum　50,103～105,159
standard deviation　99
standard enthalpy changes　41
standard electrode potential　96
standard enthalpy of combustion　41
standard enthalpy of formation　41
standard molar volume　28
standard reduction potential　96
standard solution　92,100
standard state　41～42,96
standard temperature and pressure (STP)　27
standardization　100
states of matter　17
steel　7,156,179
step reaction polymerization　106
stereoregular polymer　109,178
stoichiometric amount　36
stoichiometric (equivalent) point　95
stoichiometry　32,35,37,118
storage batteries (accumulators, secondary cells)　96
strong electrolytes　82
structural isomers　80
subatomic particle　16
sublimation　43～44,90
substitution reactions　86
substrate　63,116,135～137
supercooled　90
superheated　90
supersaturated solution　92
surface tension　90
surroundings　40～41,43
syndiotactic polymer　109
synthetic fiber　109
synthetic plastics　109
synthetic rubber　109,177～178
system　40
tacticity　109
termolecular reaction　93
tertiary structure (3D arrangement of residues)　121
theoretical density　91
theoretical yield　36～37
theory　6～8,26～27
thermochemical equation　41～42
thermochemistry　40
thermodynamics　40,127,179
thermonuclear reaction　81
thermoplastic polymer　106,114
thermoset polymer　106
threonine (Thr)　117
titration curve　95
titration error　100
titration　95,100～102,162
transition elements　77,79,175
transition state, activated complex　93
transuranium elements　77
tridentate　84
triple point　90
tyrosine (Tyr)　117

ultraviolet (UV) spectroscopy　104
unit cell　91,104,111~112
unsaturated hydrocarbons　46,86
unsaturated solution　92
valence bond theory　79
valence electrons　77~78,141
valine (Val)　117
van der Waals forces　80,94
van der Waals radii　80,162
vapor pressure lowering　92
vapor pressure　29,90,92,158
vaporization　43~44,80,90,129
variables　10,26,28
viscosity　90,109,113,185
volatile　26
volatilization method　99
voltaic cell　95
volumetric analysis　98,100~101
volumetric flasks　101
volumetric method　99,101
vulcanization　107
water softening　82
wave number　75
wavelength　16,75,81,103~104
weak electrolytes　82
weight percent　92
weight　11~13,92,98
work　40~41
α decay　81
α-particle　16
β decay　81
γ decay　81
π bonds　79
σ bonds　79
cis isomers　80
cis-trans isomerism(geometric isomerism)　80
d^{10} configuration　79
trans isomers　80